ONE YEAR OF HELL

The Heart-Breaking True Story of the 1880

Seaham Colliery Disaster

by Fred Cooper

COPYRIGHT

The right of Fred Cooper to be identified as the Author of the work has been asserted by him in accordance with the Copyright, Designs and Patents Act 1988. All rights reserved. No part of this publication may be reproduced, stored in a retrieval system or transmitted in any form without the prior written consent of the Author, nor be otherwise circulated in any form of binding or cover other than that in which it is published and without a similar condition being imposed on the subsequent purchaser.

Copyright © 2019 Fred Cooper

ISBN 978-1-9160174-0-5

TABLE OF CONTENTS

	Page
About this book	1
Acknowledgements	3
Chapter 1 – The sinking of Seaton and Seaham Collieries	7
Chapter 2 – Why the "Hell-Fire" pit	16
Chapter 3 – The poor trapper boy – The 1852 explosion	19
Chapter 4 – Two pits become one	25
Chapter 5 – The elusive 1864 explosion	29
Chapter 6 – The curse of the curve – The 1871 explosion	33
Chapter 7 – The daily work at Seaham Colliery	40
Chapter 8 – One year of hell – The 1880 explosion	52
Chapter 9 – The loss of life is much greater than first thought	60
Chapter 10 – The first inquest is opened	67
Chapter 11 – The burials begin	69
Chapter 12 – The winner of the Queen's gold cup is buried	73
Chapter 13 – The adjourned first inquest is re-opened	79
Chapter 14 – Exploration continues into No. 1 pit	83

Chapter		Page
Chapter 15 – Messages from the dead		87
Chapter 16 – The Relief Funds		92
Chapter 17 – Explorers find more messages from the dead		102
Chapter 18 – The Maudlin seam is on fire		111
Chapter 19 – The adjourned first inquest resumes		115
Chapter 20 – The strike begins		120
Chapter 21 – The adjourned first inquest resumes again		124
Chapter 22 – The widows appeal to the Relief Committee		130
Chapter 23 – The strike is causing real deprivation		134
Chapter 24 – Striking miners assault blacklegs		137
Chapter 25 – Five striking miners charged with assault		140
Chapter 26 – Mass meeting of miners from other pits		143
Chapter 27 – The strike is resolved		147
Chapter 28 – The fate of the "sacrificed men"		150
Chapter 29 – The verdict of the inquest is declared		160
Chapter 30 – The Maudlin seam is re-opened		171
Chapter 31 – The bodies of the missing men are found		176
Chapter 32 – The final inquest		180
Chapter 33 – The Durham Miners Association		182

Chapter 34 – One Year of Hell	189
Chapter 35 – In Memoriam	192
Pit Terminology & Glossary	200
Appendix 1 – The N&D Miners Permanent Relief Fund	206
Appendix 2 – Details of the bodies recovered and interred	212
Bibliography and References	219
Other books by the author	221

ABOUT THIS BOOK

Coal mining has been an essential part of British industry since Roman times. Before nationalization of the coal industry in 1947 the ownership of coal mines in Britain was fragmented and each colliery or group of collieries were under the management of a small but powerful group of coal owners. In general, the coal owners, with a few exceptions, were motivated by profit more than by social responsibility. The prosperity of the British nation has always been sustained by the labours of the coalminer. When the price of coal was high the coal owner produced more and made higher profits but when the market price fell the coal owner reduced wages or laid men off. Throughout the 19th century the British claim to be the leading industrial nation of the developed world was fuelled by the coal dug out of the earth by the coalminer.

The British coalminer was accustomed to many hardships in his day-to-day toil beneath the earth. He and his family endured the effects of strikes and industrial action fighting for his rights to a fair days work for a fair days pay. Danger from roof falls, injury from heavy machinery, fear of explosions and industrial diseases such as emphysema and pneumoconiosis became a constant partner in the coal miner's life. Their work and conditions down the mine could be described as a battle with nature.

Coal mining had always been sensitive to the economic barometer of the UK economy as it swung from periods of prosperity to depression. From the start of the industrial revolution whole communities had to be prepared for upheaval when collieries closed or workers were laid off. In times of great need the British miner knuckled down and pulled out all of the stops to improve productivity. During World War Two many young men, some reluctant and some willing Bevan Boys, did their bit for King and Country down the coalmines.

In County Durham the coalmines dominated the people in the same way they dominated and scarred the landscape. The coalminer had a unique culture and language developed through their hard lives and experiences in the mining community. Many colliery lodges had their own brass bands and miners were typified by their hobbies and pastimes of green bowls, pigeon "fancying", leek and vegetable growing, flower shows, greyhound or whippet racing - all "fresh air" pursuits - and of course by their mode of dress with waistcoats, silk scarves and the traditional flat cap. Some took up painting as a hobby such as the Pitmen Painters of Ashington and many attended College to improve their position in life for their wives and children.

In some ways the life of a coal miner could be described as a great adventure overcoming hardships and adversity. But sometimes incredibly terrible things happen down the mine. This book re-tells the horrors of six explosions underground in the first thirty years of its existence. The last explosion became known as the Seaham Colliery Disaster. On 8th September 1880 an explosion rocked the pit and the pit village of Seaham Colliery. The disaster caused the death of 164 men and boys at Seaham Colliery and threw the grieving mining community into industrial unrest for more than a year after the explosion. The story of the disaster has been told many times by local historians. This book brings together all previously known published literature and together with new research it explores and comments on the events of the first five explosions. It then takes the reader through the sixth explosion – The 1880 Seaham Colliery Disaster - one year of hell, torment and suffering in Seaham from 8th September 1880 until the final body was recovered from the mine exactly one year later.

ACKNOWLEDGEMENTS

The research and writing of a book that tells the story of One Year of Hell – The 1880 Seaham Colliery Disaster has taken me a lifetime. It has not filled my life and it has not been my sole preoccupation. To the contrary I have filled my life with my family, my career as Finance Director at The University of Sunderland and, since retirement, writing about the history of Seaham.

When I say it has taken me a lifetime, I mean it has been lingering in the back of my mind throughout my life and every now and then something happens to stir my thoughts reminding me of those horrific events in my hometown. It could be an event, a smell, a photograph, a childhood memory or a story that brings the Hell Fire Pit back to my mind. My family moved from Seaham Harbour to Seaham Colliery when I was just a boy. Every weekend many of my dads' friends and neighbours could be found in our front room sitting on a cracket in front of the coal fire talking about work down the pit while dad cut their hair with hand sheers and scissors. The favour was reciprocated when dad needed his hair cut. Why pay a barber if you can do it yourself? A pair of boots lasted an eternity when I was a boy. If they needed a new sole or heel dad would buy a sheet of leather; cut and trim the shape of the sole or heel; nail the new leather to the boot using a cobbler's last; wax the newly cut leather and put in a few "segs" that made a clip clop sound on the school assembly hall floor. Job done – another expense spared. Although the pithead baths at the Knack Pit opened in 1951 many of the older miners were accustomed to coming off shift and walking home black and grimy, covered in coal dust and with their pit clothes on. This was scary to a three-year-old. The sight of a group of black faced pitmen walking home past our front gate quickly had me running indoors. As the years rolled by the numbers walking home "black" dwindled as most miners saw the benefits of taking a hot shower at the new pithead baths rather than washing in the tin bath at home in front of the fire. From my bedroom window on an evening I could see and hear the work going on at the pithead only four hundred yards from our house.

The silhouette of the High Pit dominated the view from Eastlea Road

The pit yard was my playground during the day and the washery and screens were lit up throughout the night. The sight and sounds of the tubs leaving the cage at the pithead and clanking and banging as they progressed up the gantry on their way to the washery became a familiar noise in my ears as I went to sleep. My thoughts often lingered on the work of the pitmen employed in every conceivable skill and trade who laboured to win that coal and bring it to the surface. Power loaders, stonemen, development workers, fitters, electricians, blacksmiths, joiners, onsetters and many others all worked underground in teams winning the coal. On bank were the washery and screen hands, labourers, maintenance shop men, steam engine shunters and railway workers employed to escort the coal wagons down the mile long inclined railway to the docks.

My father worked underground. He was an experienced and hardworking collier who was always concerned for others before himself. I could only have been seven or eight years old when the front door knocker rapped and mother answered the door to find an ambulance driver stood on the step. She dashed out to the ambulance to find Dad lying on a stretcher. An endless haulage wire rope had snapped and the whiplash had caught him on the back and neck. Despite being in tremendous pain he had insisted that the ambulance driver brought him home so he could reassure mam that everything was alright before they took him to hospital. He constantly reminded me from a very early stage that he would never allow me to work down the mine. This was a far cry from the early days of the 19th and 20th century at Seaham Colliery when it was accepted that your dad spoke to the overman and you were set on to work down the pit the week after you left school. Well – I was set on to start work at Seaham Colliery three weeks after leaving school. To my dad's relief and delight I began my long career in financial management in the newly built Wages Office just along from the lamp cabin and next to the time office. So, I was brought up, educated, worked and experienced life in a colliery village amongst family and friends most of whom worked at and down the pit. I knew about the Seaham Colliery disasters. I was aware of the dangers of working in the mine. Amongst other duties as a junior in the wages office I had to fill in miners claim forms for industrial injury benefit (Form BI82). Details were taken from accident report forms made out by the Deputy. Although there were no major disasters during my time at Seaham Colliery the accident report forms made explicit the dreadful injuries to miners arising from their day-to-day work.

In re-telling the story of the "One Year of Hell" I have to acknowledge all of the men and women I grew up with in the colliery village; my school friends; my work colleagues in the 1960's and 1970's and my family. All of these people shaped my views on life in general and the way of life of an extraordinary social and cultural group – the coalminers of Seaham Colliery. I begin the book with a general description of the Durham coalfield and the building and development of the town of Seaham and the pit village of Seaham Colliery. For those readers who are unaccustomed to life in a coal mining community, pitmatic terms and phrases I have provided a glossary of coalmining names, terms and descriptions at the end of the book which may help in understanding the events of the disasters as they unfold.

There are a number of individuals and groups who have provided photographs, information, facts and personal experiences and reminiscences which have enriched the content of this book and for which I am extremely grateful. For any contributors that I have overlooked to mention I do apologise.

My sincere thanks for your assistance are extended to: -

Linda and Bill Baker, Seaham Family History Group
Alan Charlton, Seaham Family History Group
Brian Scollen, East Durham Heritage & Lifeboat Group

Fred Cooper BSc ACMA CGMA

CHAPTER 1

The Durham coalfield and the sinking of Seaton and Seaham Collieries

The presence of coal beneath the surface of County Durham had been known since Roman times. Through many centuries the process of mining for coal has developed into an industry. In many districts where coal was near the surface it was collected and used in blacksmiths furnaces but in medieval times by far the greater number of our ancestors preferred charcoal for domestic purposes. However, attitudes began to change from the 16th century and the domestic and foreign demand for coal prompted greater and deeper exploration into the Durham coalfield.

Coal in the County Durham coalfield occurs in seams of various thicknesses and quality and at a range of depths up to 2,000 feet below the surface. Not all of the seams are workable. In the west of Durham some of the coal seams outcrop on the surface and, covered by clay, sand or gravel, they can be easily mined whilst towards the Durham coast the seams extend far below and under the sea and can only be reached by sinking deep shafts. As a consequence, communities were first populated and coal mining activity began first in the west of Durham. Gradually shaft sinking technology and know-how advanced and the problems of penetrating deeper into the earth were overcome and coal owners began financing deep coal mines nearer the coast.

Mining engineers found that once a coal seam was reached it was not a straightforward process to follow the coal seam until it was worked out. In many places in the Durham coalfield there are geological faults where the seam

Diagram 1 (Not to scale)

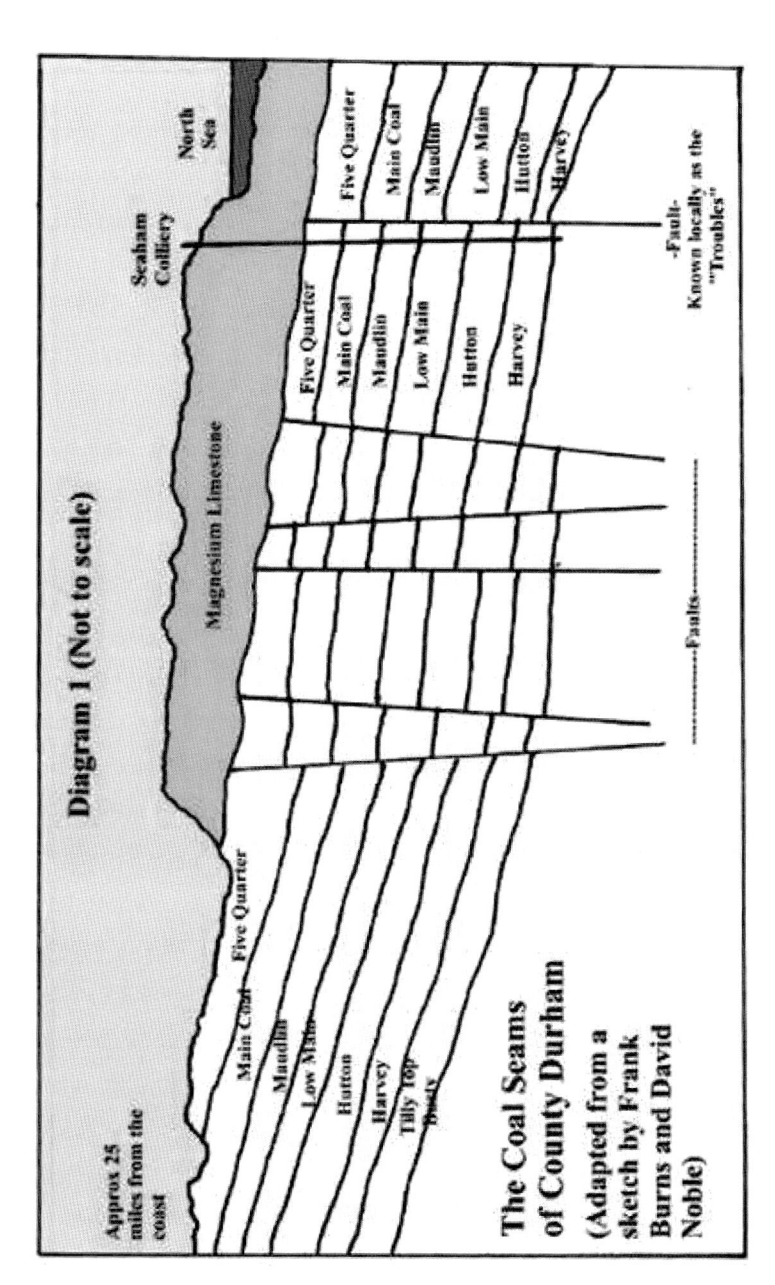

Approx 25 miles from the coast

Main Coal — Five Quarter
Maudlin
Low Main
Hutton
Harvey
Tilly Top
Busty

Magnesium Limestone

Seaham Colliery

North Sea

Five Quarter
Main Coal
Maudlin
Low Main
Hutton
Harvey

Five Quarter
Main Coal
Maudlin
Low Main
Hutton
Harvey

-----Faults-----

-Fault-
Known locally as the "Troubles"

The Coal Seams of County Durham

(Adapted from a sketch by Frank Burns and David Noble)

drops or rises and consequently the miner has to find a way through the fault back to a workable coal seam. These faults in many cases can cause the coal seams to dip and rise at very steep angles and if not overcome can lead to the closure of the mine (DCEC Group, 1993). Diagram 1 above shows the various coal seams in the Durham coalfield with the common names used by the Durham miners. The main geological faults can be seen on the sketch. A fault in the workings at Seaham Colliery a little to the east of the downcast shaft (No. 1 and 2 Pit) was sometimes locally referred to as the "troubles".

The width of the coal seams varied from 1 foot in the west of Durham to 4 feet towards the coast and in some exceptional cases up to 8 feet. At Seaham the middle seams i.e., the Main Coal (5 feet thick), Maudlin (4 feet thick), Low Main (4 feet thick) and Hutton (4 feet thick) were the most productive. At the coast the Main Coal seam generally lays at 1,400 feet; the Maudlin seam at 1,500 feet; the Low Main seam at 1,550 feet and the Hutton seam at 1,600 feet below the surface. At Seaham Colliery the depth of each seam including the depth beyond the fault was as follows: -.

1. Main Coal Seam .... 1,310 feet   F

2. Maudlin Seam .... 1,370 feet   A

3. Low Main Seam .... 1,390 feet   U ... Main Coal

4. Hutton Seam .... 1,530 feet   L ... Maudlin & Low Main

5. Harvey Seam .... 1,690 feet   T ... Hutton Seam

6. Busty (Downcast shaft bottom)... 1,800 feet   .

Diagram of levels at Seaham Colliery on 8th September 1880

The creation of Seaham Harbour and the pit community at Seaham Colliery is the most remarkable of any town in the Durham coalfield. In 1808 Seaham was, according to a diarist "Memoirs of a Highland Lady" (Smith, 1898) a most primitive hamlet, a dozen or so cottages; no trade; no manufacture; no business; residents were mostly the servants of Sir Ralph Milbanke and apart from the Clergyman's family there were none of the gentler degree. The most memorable event to occur at Seaham was the marriage of Lord George Byron on January 2nd, 1815 to Miss Isabella Milbanke at Seaham House. Her father Sir Ralph Milbanke was the first to envisage a bustling harbour at Dalden Ness just half a mile from his manor house at Old Seaham. Plans were drawn up for "Port Milbanke" but his ambitions were abandoned when Sir Ralph fell into financial difficulties because of his constant electioneering as a Member of Parliament in addition to the payment of a dowry to Lord Byron upon the marriage of his daughter. In 1820 the port and town still did not exist. The 3rd Marquess of Londonderry who bought the twin estates of Seaham and Dalton in 1821 knew of the proposal and decided that a port at Dalden Ness would be ideal for shipping coals from his wife's' mines in Pittington and Rainton. Building of the harbour was begun by the eminent engineer William Chapman on a lonely and uninhabited part of the coast and on 28th November 1828 the port and town of Seaham Harbour was officially inaugurated. The famous Newcastle architect, John Dobson, was commissioned to design the town which originally was planned around a main street flanked by two crescents facing the sea. Three classes of houses would be constructed. The first-class houses in the South Crescent would have six rooms; the second-class houses in the North Crescent would have four rooms and the third-class houses would be cottages flanking each side of the railway line leading down into the docks. Although the Marquess was pleased with the plans, he did not have enough financial backing to proceed with such grand proposals. He chose instead to lease land to individual builders for shops, industrial premises and domestic housing and so the ambitious ideas and uniformity of construction in Dobson's plan was lost. The local workforce was inadequate to satisfy the demand for labour in the town and the port and a flood of immigrants arrived. Within thirty years the port could accommodate three hundred ships and the population of the town had grown to exceed 8,000 persons.

Coal mining at Seaham began with the sinking of Seaton Colliery by the Hetton Coal Company in 1844. It was owned by The Earl of Durham and later known as "The High Pit". Two staples of six feet diameter were first put down on 31st July 1844 and the actual sinking commenced on 12th August 1845. It was feared that problems would be found with sinking through water bearing strata similar to that found during the sinking of Murton Colliery six years earlier. The difficulties encountered in the sinking of Murton Colliery pushed costs up to £80,000 (Fordyce, 1860) as two high-pressure engines of 450 horse power supplied with steam from eighteen cylindrical boilers and flanked by two chimneys each eighty feet high were needed to drain the incessant flow of water into the shaft. Although the sinkers worked in appalling conditions at Seaton Colliery to everyone's relief the difficulties were overcome and the water was adequately pumped out as the shaft progressed with a diameter of 14 feet and to a depth of 1,839 feet.

By the latter half of the 1840's the best seams of coal at pits belonging to The Marchioness of Londonderry at Rainton and Pittington were becoming exhausted. Consequently, the sinking of a second pit at Seaham by the 3rd Marquess of Londonderry to be called Seaham Colliery began on 13th April 1949 only 150 yards apart from the shaft of Seaton Colliery. Using the same equipment, the sinkers constructed a shaft 14 feet in diameter and to a depth of 1,797 feet. Three months after the start of sinking operations in July 1849 the sinkers discovered a toad embedded in the limestone 183 feet from the surface. It was sent to bank alive but died a few minutes afterwards. It was given to Tommy Chilton of the Mill Inn who was known as Nicky Knack because of the collection of curiosities that he displayed at the Inn. The colliery at Seaham was soon to take on that nickname of the "Knack Pit". Both of the sinking operations at Seaham were managed by Mr Nicholas Wood, later to become President of the North of England Institute of Mining and Mechanical Engineers, and John Buddle the renowned Coal Mining Engineer and Colliery Viewer who was later succeeded by Mr George Hunter. No pumping engines were erected but the two 150-horse power winding engines were adapted for pumping by attaching a pumping beam 36 feet long to work a series of pumps with a 19-inch working barrel (Fordyce, 1860).

By early 1852 sinking operations were fully completed at Seaham Colliery and on 22nd May the Newcastle Journal carried an advertisement offering for sale by public auction all of the materials and equipment employed in the sinking and winning of the two collieries. Almost three years after sinking operations began the first coal at Seaham Colliery later known as "The Low Pit" was drawn on 27th March 1852.

Every colliery shaft sunk in the Durham coalfield promoted the growth of a settlement around the mine composed entirely of the coalmining community. Surrounding the two pits of Seaton and Seaham Colliery twenty-one rows of miners' houses were built between the 1850's and the 1890's to accommodate over 800 miners and their families. Tenders for the first of the colliery houses were advertised in the Newcastle Courant on 15th August 1851. The advertisement read "To Contractors: Tender to be let for the building of eighty pitmens' cottages at Seaton Colliery. Plans will be deposited with Mr William Coulson at the Colliery Office who will give any other information that may be required". Many of the colliery houses were built with little regard to the necessities of the people who had to occupy them. Streets were unpaved and undrained; water was provided by a communal standpipe and toilet facilities consisted of an outside "privy-midden" where coal ash from the coal fire was thrown in on one side whilst the other side served as a latrine. In the absence of a proper sewage system the coal ash was thrown down the latrine to mask the smell. The privy-middens were cleaned out regularly onto hand carts by men employed by the local health board and the contents were incinerated or composted into fertiliser. The houses were divided into three classes. The first-class houses were usually allocated to the colliery officials and had two rooms on the ground floor with a kind of loft above that was reached by a ladder. The next class had only one room below with a loft above while the third class had only one room on the ground level.

Butcher Street, Seaham Colliery (Built c1850) - Photographed from the top of the pit heap
Courtesy of Madeline Eggleston

The colliery houses at Seaham Colliery were built and owned by Lord Londonderry the coal owner. In some of the other colliery districts the houses were built by private builders who would lease each of the houses to the coal owner for £3 or £4 per year. The miner occupied the house rent free for as long as he was employed at the pit. Some of the pit houses had a reasonable size garden and much of the land surrounding the pit village was used as allotments or kitchen gardens to grow vegetables. Colliery houses were often used as an incentive to induce men, especially those with large families of sons, to work for a coal owner at a particular mine. In 1883 Mr Patterson, Assistant Manager to the Marquess of Londonderry travelled to Hamilton in Scotland to recruit miners for Seaham Colliery because of the shortage of men in Seaham. The Hamilton miners were offered five shillings a day with free house and coal and their travel costs paid to Seaham Colliery.

The nature of mining made the pitmen and their communities fairly distinct from other labouring classes and they were often regarded by outsiders as "clannish" although it could be said that the community were merely looking after the interests of their own families and workmates (Fowler, 1982). The daily routine of a pitman's wife was very different from that of other women. Consider a common enough situation in a pitman's family. The men in the household could be employed down the mine as a hewer, putter or a shiftman. Her eldest son could be working at the coal face as a hewer; her youngest son could be a putter pushing the tubs from the coal face to the landing and her husband could be a shiftman repairing the roadways. Each one worked a different shift pattern and went to work and returned home at different times. She had to ensure there was ample hot water for each of them to bathe after their shift; she had to mend and wash their clothes ready for the next shift and she had to prepare a hot meal for each of them before they went to bed. In addition, she had to keep everything neat and tidy in the home as well as find time to feed herself and to sleep. Under ordinary circumstances she fed her family on good basic cooking. Considering the amount of physical exertion required down the mine a pitman's wife had to make substantial meals for the menfolk. One of the favourite meals in a pitman's house was suet pudding with beef and kidney filling, cooked in a basin covered with a clean cloth firmly tied and then boiled in a pan of water for three to four hours. That would be followed by a sweet suet pudding covered with treacle or spotted dick with sultanas and raisons. Often as a treat for supper she would prepare tripe and onions poached in milk. (Whellan, 1894)

As the population of Seaham Colliery expanded a new church was needed to fulfil the spiritual needs of the community as the existing parish church of St Marys was more than two miles away at Old Seaham. A new church called Christchurch was built and opened in 1857 across the road from the pit entrance to Seaton and Seaham Colliery. It was at first created as a chapelry of St Marys until the parish of "New Seaham" was created in 1861 and Christchurch became the parish church.

The countryside where once there were green fields and arable farmland was transformed from the 1850's into a mighty industrial complex by the erection of pit-head buildings, lofty steam-engines, chimneys belching columns of smoke into the sky, noisy blacksmiths' shops, wagon ways, grimy coal screens, slag heaps and row upon row of colliery houses bristling with miners going to or coming from their work at the pits.

CHAPTER 2

Why the "Hell Fire" Pit

Temperatures down a pit can in places rise to 80 degrees Fahrenheit but that alone does not warrant the fearful name given by the newspapers and the Seaham Colliery miners to their workplace. The Hell Fire pit!

No – it was more than that. For thirty years since it first drew coal there were dreadful accidents. That was nothing unusual for coal mines in Victorian times. The wives and families of men working down the pit reluctantly accepted the day-to-day accidents that injured and maimed their husbands as another burden to bear and more hardship to overcome. But in addition, came the terror of miners the world over – explosion. In less than thirty years since coal production began at the two collieries there were six horrific explosions that caused great injury and loss of life. The explosion record of Seaham Colliery can be summarised as follows: -

Early 1852 – Two explosions - No lives lost

16th June 1852 (Wednesday) – Explosion – 6 lives lost

6th April 1864 (Wednesday) – Explosion – 2 lives lost

25th October 1871 (Wednesday) – Explosion – 26 lives lost

8th September 1880 (Wednesday) – Explosion- 164 lives lost

Note: The newspapers initially reported an explosion at Seaham Colliery on 20th September 1872 with no loss of life. Upon investigation it was found that it was far less serious. Some paper and dust had ignited beside a shotfirer who had just completed two shots in a stone drift in Number 3 Pit. The fire was extinguished by another miner in close proximity and there was no explosion.

The causes of explosion in a coal mine can primarily be attributed to poor or defective ventilation; carelessness of workmen in using naked lights or shotfiring and the negligent management of the mine. The ignition of explosive gases has the effect of not only spreading destruction in the locality in which it occurs but it also extends its appalling effects to distant parts of the mine. The subject of pit explosions engaged the attention of both Houses of Parliament and the government was eventually induced to appoint four independent and able scientific men as Inspector of Coal Mines in 1850. By 1906 there were twelve Inspectors and twenty-six Assistant Inspectors. Inspectors were mining experts with the power to inspect mines above and below ground for all matters related to safety and for compliance with appropriate legislation. They had to be notified of all fatal accidents and would gather various statistics. Inspection could be following an accident, on invitation or following a complaint from the miners. As the number of inspectors increased over the years routine, unplanned inspections began and they became the dominant part of their work.

Each inspector was responsible for a geographical region and produced a report for that region. The first four inspectors covered all of Great Britain between them, losing areas as further inspectors were appointed until by 1855 there were 12 regions or districts. The numbers of districts varied over the years between 6 and 12 (with additional ones for metalliferous mines from 1873-1901) as restructuring took place, or merging of districts was needed because of staff retirement or death. Mr Matthias Dunn was responsible for the Durham District in 1850 and he attended all cases of serious accidents and explosions in collieries in the Durham district.

The first five explosions at Seaham Colliery are covered in the next few chapters. The full story of the 1880 explosion will be covered in more detail from Chapter 8.

CHAPTER 3

The poor trapper boy
The 1852 explosion

An unlucky co-incidence links all of the explosions at Seaton and Seaham Collieries. Every explosion happened on a Wednesday. On 16th June 1852 a dreadful explosion of fire-damp took place at Seaton Colliery. Although the pit had only been in production for five months it was being described as a fiery pit. Two explosions from fire-damp had already taken place although without loss of life. This time the men were not as fortunate. The explosion took place about noon and shortly afterwards several men volunteered their services and descended the pit. They found the pitmen and boys nearest to the shaft quite safe. They tried to explore further inbye but the afterdamp was so strong they could not go further and they returned to bank. Soon afterwards a second rescue party went down and were able to explore further. On reaching the location of a trap-door they found the body of ten-year-old trapper boy Charles Halliday. The door had been blown off its hinges and the poor boy had been thrown about thirty yards by the force of the explosion against a wall and was partly buried by a roof fall of stones. It transpired that the older brother of Charles Halliday had heard the explosion and ran past his poor brother's body lying partly hidden under the fall of stone, dashed through thirty yards of fire-damp, and escaped unhurt. The exploring party progressed onward finding the body of a horse and then the body of William Simpson who was partially burnt. They then found the bodies of another four men who were not in the least burnt and who must have died from the effects of the afterdamp.

In total six miners were killed and several others were injured. The names of the deceased were: -

- Charles Halliday aged 10
- William Simpson aged 27
- John Simpson aged 36
- Andrew Simpson aged 18
- John Defty aged 53
- John Pratt aged 20

John Defty left a grieving widow and nine children. In early Victorian times colliery managers often came under pressure from miners to employ their children, both girls and boys with some as young as eight years old. After a serious accident at Huskar Colliery in 1838 in which the mine was flooded it was revealed that eleven girls aged from eight to sixteen and fifteen boys aged between nine and twelve had perished. The story of the accident appeared in London newspapers and Queen Victoria who read the reports put pressure on her Prime Minister, Lord Melbourne, to hold an enquiry into the working conditions in Britain's' factories and mines. The Children's Employment Commission published its first report on mines in 1842 which legislated that women were banned from working in the mines and that no boys under the age of ten could work underground. Charles Vane, 3rd Marquess of Londonderry led the opposition to the Bill in the House of Lords and declared that some seams of coal required the employment of children and that certain pits could not afford to pay men to carry out such work and would close down. In 1872 the age of boys who could work in the mines was raised to twelve and then eventually to thirteen in 1903. Ten-year-old Charles Halliday was a trapper boy. His duties were to open and close a door which kept the air supply flowing in a particular direction for the men in that seam. Failure to close the door promptly and securely would result in a break in the ventilation process through the workings and the possibility of a build-up of fire-damp gases. It was an eerie job; all alone for ten hours with a faint flicker of light from a candle or lamp interspersed with putters pushing their coal tubs back and forth to the shaft. They were often subject to beatings from the putters if they fell asleep and did not open the doors promptly. Poor Charles Halliday's hard and miserable existence was tragically cut short before he had a chance to experience any joy in his life. No childhood games in the street for Charles or running through fields with friends. Charles was buried on 19th June 1852 in the ancient church of St Mary's at Seaham.

The Trapper Boy

The bodies of the six men and boys who died at midday in the explosion on that fearful Wednesday were brought to bank that afternoon. News of the explosion had spread all around the district and wives, children and families congregated waiting for news and it was said that the scene at the pithead was heart rendering. All of the survivors and several of the exploring party who were affected by the afterdamp were attended to by Mr Ward, surgeon of Seaham Harbour when they returned to the surface.

The inquest was held in the Mill Inn on 23rd June before the Coroner Mr TC Maynard. Mr Matthias Dunn the Government Inspector of Mines was in attendance in addition to Mr Morton the agent to the Earl of Durham and viewers and mining engineers from many other collieries in the region. Ralph Dunn a coal hewer gave evidence. He told the inquest that on the day of the explosion he was working with candles and blasting coal with powder in the first South-west drift. There was plenty of air ventilating his workplace which was about six feet wide. The explosion took place about 500 yards away. James Bruce, a Deputy Overman, stated he was working at the west side of the shaft and upon hearing the explosion went into the North side and then returned again. He passed the body of young Halliday and the bodies of the other men which were beyond the cross-cut. The body of Simpson was about fifty yards beyond the cross-cut and that of John Defty and John Pratt were about twenty yards further up. He found the second Simpson in the back-drift about two or three yards from the fall of stone. The third Simpson was found at the stenton about sixty yards from his workplace. These men had been working with candles. No one had detected any gas or any sign of danger before the explosion and no gas was detected after the explosion. Two colliery viewers gave evidence to the inquest. Mr Forster and Mr George Elliott had both visited the pit the previous Monday. Mr Forster concluded after looking at the face of the drift that the coal in that quarter was tender and he had no doubt that as the coal was very soft that gas would escape very readily. Mr Elliott confirmed his opinion that the explosion happened in the back west drift and that they had found the candle at the end of the brattice about five feet from the coal face. The Coroner, Mr Maynard, heard evidence from a number of other men and from the management. He concluded that the explosion had occurred in the back-drift. It was probable that a "blower" of gas had been released through a fissure in the coal face and this had been ignited by the open flame from a candle causing the explosion. Mr Dunn, Inspector of Mines, was more direct in his summing up of the method of working. He added that although there was no indication of gas prior to the explosion and that everyone working in that district considered that the ventilation was adequate it would have been prudent if the management had issued Davy safety lamps to the men when powder was being used. He also agreed that the workings at the pit were still under development as production had only recently commenced. However, in his opinion an increase in the width of the drift from six feet to eight feet would improve the ventilation and the volume of air to the workplaces and would make much safer working places for the hewers and the shotfirers. The jury retired to consider the evidence and upon returning the Coroner announced their verdict that the explosion was "purely accidental".

It is ironic that something good should come out of such a tragic event. The inquest held on 23rd June 1852 in the Mill Inn closed. The jury and reporters dispersed. Discussions of a most interesting nature then took place amongst the viewers and other scientific men who had attended the inquest. Amongst those present were Mr TE Forster, Mr TC Maynard, Mr GB Forster, Mr H Morton, George Elliott, Mr E Sinclair; Matthias Dunn the Inspector of Mines for the Durham Area and other mining engineers. It was during these discussions in the Mill Inn that the first suggestion was made to form a Mining Institute for the North of England and the title, regulations and the constitution of this new Association was settled that evening (Volume 15, Transactions of NEIME). These proposals were agreed at a subsequent meeting on 3rd July and then formally adopted at a meeting in the Coal Trade Office, Newcastle on 31st July 1852. Those present were concerned coal owners and mining engineers interested in the prevention of accidents in mines and the advancement of mining science. The North of England Institute of Mining and Mechanical Engineers was now formally created. At a further meeting of its 70 members held on 21st August 1852, Nicholas Wood was elected President; four Vice-Presidents and twelve Council members were elected. The object of the Institution was twofold – first, by the concentration of professional experience to devise measures which might avert or alleviate the dreadful calamities which had produced such destruction to life and property; and secondly, to establish a literary institution applicable to the theory, art and practice of mining.

The advancement of the science of mining would play an increasing role in working practices in the Durham mines and make a real impact on the safety of pitmen throughout Northumberland and Durham and it started in the Mill Inn, Seaham.

A blue plaque commemorates this momentous occasion above the door of the Mill Inn.

CHAPTER 4

Two pits become one

The sinking of pit shafts was an expensive and risky business with no guarantee of a profitable outcome. The costs could range from tens of thousands of pounds to upwards of £80,000 if geological difficulties were encountered such as at Murton and Wearmouth collieries. As a consequence, almost every colliery was built with one shaft - one way down and only one way out.

Both coal owners at Seaham - The Marchioness of Londonderry and the Hetton Coal Company - were aware of the accident at Page Bank Colliery on 1st October 1858 when the shaft caught fire and ten men and boys suffocated by inhaling the smoke fumes coming down the shaft. Discussions took place between the two coal owners about the potential for such an accident happening in the shaft at each of their pits. The Londonderry's not only owned Seaham Colliery but were also shareholders in the Hetton Coal Company that owned Seaton Colliery. Agreement was reached on the purchase of Seaton Colliery by the Marchioness of Londonderry in August 1860 thereby uniting the ownership of both pits.

More shaft accidents happened soon afterwards. One of the most appalling and heartrending catastrophes that has ever occurred in Britain took place on 16th January 1862 in Hartley New Pit, near Seaton Delaval. Over the mouth of the pit was the beam of a pumping engine, the largest and most powerful in the North of England, the beam weighing about forty tons. The men were being drawn up in the cage by means of the winding machine when the beam of the engine broke and fell into the pit, meeting, in its downward course, the ascending cage carrying eight men to bank, the enormous mass crushing everything in its way. Five were killed instantly, and three were afterwards rescued alive. The beam struck the top of the brattice with such violence that the whole of the massive wooden and iron framework was hurled to the bottom of the mine, thus cutting off all means of escape from the lower portion of the mine, in which 204 men and boys gasped in the foul air until it was exhausted completely and they all perished (Fordyce, 1867).

The Hartley Colliery Disaster Inquest on 6 February 1862 returned a verdict of 'accidental death'. The jury expressed their strong opinion of the imperative necessity that all working collieries should have at least a second shaft or outlet, to afford the workmen the means of escape should any obstruction take place.

However, one prominent mining engineer gave his opinion that "Parliament should pass an act this session" requiring two shafts at every colliery. On 7th August 1862, just 6 months after the inquest and less than 7 months from the disaster, an Act of Parliament (the Act to Amend the Law Relating to Coal Mines of 1862) was passed. This required all new mines to have two shafts and all existing mines to ensure access to a second shaft before the end of 1864.

The owner of the united Seaton and Seaham Collieries, Lady Frances Ann Vane, was fortunate. The two pit shafts were only 150 yards apart and access could easily be made underground between one shaft and the other. In anticipation of the passing of the Act an underground connecting roadway was completed between both shafts by March 1862. This was to prove to be a most fortunate act. An accident similar to that at Hartley Colliery occurred in the shaft of Seaham Colliery on 4th April 1862. About half-past eleven in the morning when there was between three and four hundred men working in the mine one of the cages coming up the shaft got out of the "skeets" (or guides) which kept the cage in position and it came into a violent collision with the cage that was descending. The shock of the collision drove the loosened cage forcibly against the brattice work which divided the shaft and about ten fathoms of the brattice fell down the shaft blocking up the shaft in much the same way as at Hartley Colliery. Luckily for the miners underground the connecting road had recently been made into Seaton Colliery and in a very short time everyone down the pit was safely at bank. It was estimated that within two hours of the shaft accident the stythe caused by the shaft blockage would have suffocated the men.

A further incident in the shaft at Seaham Colliery ten years later on 9th August 1872 underlined the importance of uniting both of the pits underground. Two tubs became detached from a train travelling on bank. The tubs hurtled down the incline into the shaft causing great damage to the timberwork and displacing the slides which guide the cages while ascending and descending the shaft. Three shaftmen named George Brown, John Purvis and Henry Tate were instructed to repair the damaged slides. Having carried out the repairs they replaced the cage in what they thought were the correct slides and, mounted on the top of the cage with three other miners inside the cage, began their ascent to bank. Shortly after leaving the bottom of the shaft they became aware that they had placed the cage in the wrong set of slides and that the descending cage was coming down the same set of slides. There was no time to signal the banksmen with the rapper wire to stop the cages and when about half way up they encountered the dreadful calamity of the downward cage hitting them knocking Brown and Purvis into the sump some 300 yards beneath them. By some miraculous means Tate was protected from the same fate by the timbering of the shaft although he was severely injured. The three men inside the cage; Wilkie Rowell, under-engineman; Robert Smurthwaite and John Jeffrey both banksman escaped without injury. Once again, the underground roadway to the other shaft led the men already down the mine to safety.

Miners at the jointly owned Seaham and Seaton Colliery could confidently descend the pit from 1862 in the knowledge that if there were to be a shaft accident similar to Page Bank Colliery or Hartley Colliery they had a ready means to escape back to bank through the adjoining road to the other shaft.

CHAPTER 5

The elusive 1864 explosion

At the previously named Seaton Colliery - now combined with Seaham Colliery - an explosion of fire-damp occurred on the afternoon of 6th April 1864.

Work had been progressing to improve the ventilation arrangements for conducting the return air in the roadway connecting the Seaham Colliery to the Seaton Colliery shaft in, and above, the main coal seam. To do this a large arched flue was being constructed and much of the arched brickwork had been put in. Ahead of the new brickwork a large excavation was being prepared to extend the brick flue forwards. The men encountered a hitch or fault when digging out this area and the roof above fell leaving a void which was a considerable height above the level of the top of the arching. While the men were engaged in removing the debris a fall of stone occurred which was immediately followed by an explosion of fire-damp.

Four persons working directly beneath the fault Messrs Tones, Fairley, Heppell and a boy named Burdon were badly scorched by the explosion. All of the injured were attended to by Dr Beatty, the colliery surgeon, and his assistant. One of these, Tristram Heppell, who was in charge of the work and another William Fairley died some days later from the effects of the explosion. Heppell was a member of No 5 Company of the 2nd Durham (Seaham) Artillery Volunteers. His funeral procession was attended by 200 Volunteers headed by Captain Mann, and preceded by the Volunteer Band; they marched the two miles from Heppell's house to St Mary's Church to the strains of the "Dead March". Three volleys were fired with great precision by his comrades at the grave. So popular was Heppell that only one quarter of the family and friends attending could actually get inside the church.

The evidence disclosed at the inquest revealed that a large fall of stone preceded the explosion. No gas had been detected in that place before the explosion and naked lights were being used by the men employed in constructing the brick arch. It had been necessary to fire gunpowder shots to overcome a difficult stone section but before this happened gas was searched for using safety lamps and none was found. Ventilation readings were taken 150 hours before the incident and this showed a reading of about 5,000 cubic feet of air per minute in that place and again no gas had been detected during those tests. The underviewer himself told the inquest that he had used a naked light in the upper part of that place for about three hours on the morning of that day and he had not detected any symptoms or signs of gas.

The inquest found that it was probable that the gas had been released from the fault or hitch. This gas had accumulated in the upper part where the void had developed and had been forced down by the fall of stone where it met with the men's naked light causing the explosion. The jury returned a verdict of accidental death. Shortly afterwards safety lamps were adopted for working in all parts of the Seaham Colliery. (Atkinson, 1864)

This incident is described as the elusive 1864 explosion. With the exception of HM Inspector of Mines report there was only one other reference made to the incident and to the deaths of Fairley and Heppell in local newspapers. That reference was a six-line mention in the Seaham Weekly News. The Seaham Librarian John E McCutcheon, author of "Troubled Seams" published in 1955 had heard of the incident and found the tombstone of Tristram Hepple in the ancient churchyard of St Mary's, Old Seaham. (McCutcheon, 1955)

Courtesy of Seaham Family History Group

The headstone reads: -

> "In memory of Tristram Heppell who
> was burnt by an explosion of gas at
> Seaton Colliery and died on the
> 22nd April 1864, aged 50 years"

William Fairley aged 42 died from the effects of burns from the explosion and was buried in Christchurch on 13th April 1864.

It is a complete mystery why an explosion at the colliery injuring four men and eventually resulting in the death of two of them should have received so little coverage in the local newspapers. Reports of the proceedings of the inquest could also not be found in the local newspapers. It is fortunate that HM Inspectorate of Mines gave details of the explosion in their statutory report preserving the facts behind these tragic deaths for relatives and mining historians.

CHAPTER 6

The curse of the curve

The 1871 explosion

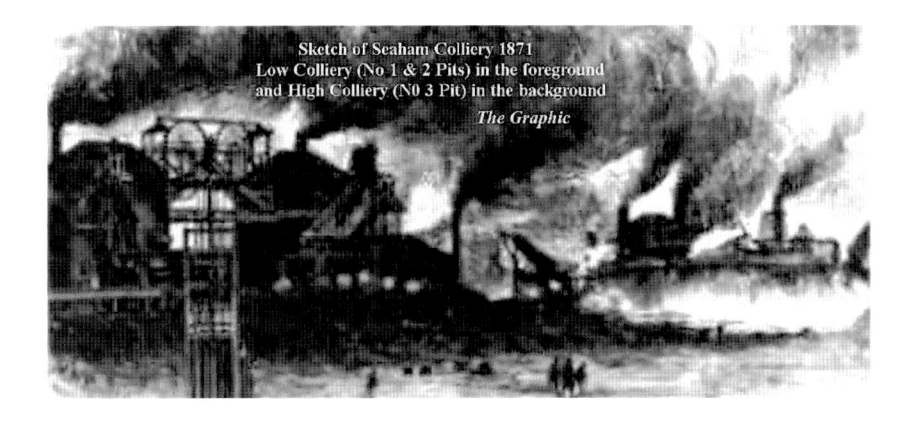

Sketch of Seaham Colliery 1871
Low Colliery (No 1 & 2 Pits) in the foreground
and High Colliery (N0 3 Pit) in the background
The Graphic

It is January 1871 and the now conjoined Seaham and Seaton Colliery has been in coal production for almost twenty years. A journalist and an artist from The Graphic visited Seaham Colliery to gain an insight into the work of a typical coal mine and the work of a coal miner in the North of England for an article published in The Graphic on 28th January. The assistant colliery viewer, Mr Thompson, provided the journalist with a description of the work carried out underground. He described Seaham Colliery as one of the largest collieries in the North of England employing about 1,100 persons and raising 1,600 tons of coal daily. The journalist witnessed the panting of engines; rushes of steam; rattle of ropes and the roar of loading and unloading of coals echoing around the colliery village throughout the day and the night. The illustration above shows a typical scene at the colliery drawn by the Graphic's artist in 1871. The assistant colliery viewer provided the journalist with details of the general working arrangements at the colliery. "There are two shafts each 14 feet in diameter. One shaft is the Low Colliery and is referred to as the Nos 1 and 2 Pit. It is divided into two equal parts with a cage in each side fixed in wooden slides. Each cage is capable of carrying four tubs of coal and the cage is designed to lift a maximum load of eight and a half tons. The other shaft is the High Colliery (previously Seaton Colliery) and is referred to as the No 3 Pit. There are three 150-horse power winding condensing engines for raising the coal from the mines. The rope roll on which the rope winds is 22 feet in diameter. At the bottom of the pit an onsetter takes the empty tubs out of the cage replacing them with full ones. Then a boy called a "driver" with a pit pony takes from six to twelve empty tubs at a time from the bottom of the shaft into the flat where he exchanges the empties for full tubs. This part of the underground road is from six to seven feet high. Here the putters – boys driving small ponies – in a height of 4 feet 6 inches take the empty tubs to the hewers who are digging coal at the coal face and they then return with full tubs to the flat.

Sketch of hewers and putters at Seaham Colliery (The Graphic 1871)

The hewer at the coal face uses a pick axe to dig the coal from the coal seam and every two feet advanced into the coal seam he places a piece of timber horizontally across the roof with an upright on either side forming a supporting framework for him to work in."

Nine months later that same journalist and artist from the Graphic re-visited Seaham Colliery but it was for a more disturbing assignment. On Wednesday 25th October 1871 there were a large number of men and boys working in the mine but only twenty-nine of them were working in the Hutton Seam workings about one mile from No. 3 pit shaft. Fourteen of those were shifters; six were hewers and nine boys. Suddenly and without any prior warning of danger an accumulation of gas happened in the Hutton Seam workings. At half-past eleven that evening the gas took fire initiating a loud violent explosion. The ferocity of the blast flattened everything along the workings and up the shaft throwing all of the working shaft gear into disarray. Large quantities of the brick lining of the shaft were thrown down the shaft and the engine house on the surface was considerably broken although fortunately the winding engine escaped damage. In the colliery village the shock of the explosion was distinctly felt. Dense smoke was issuing from the shaft. Mr Corbett, the head viewer who lived at Londonderry Dene House just over half a mile away was summoned. Mr Corbett, Mr Dakers the under viewer and Mr Thompson, assistant viewer accompanied by a deputy overman immediately descended Nos. 1 and 2 Pit shaft which was found to be undamaged by the explosion. On reaching the neighbourhood of No.3 Pit shaft they found that the explosion had not affected the main seam workings and the ten men working there were uninjured. The exploring party and the ten men advanced further into the Hutton Seam workings where they came upon the dead bodies of several miners. They began to recover what bodies they could but the "stithe" became so dense that they had to suspend further explorations and returned to bank. When they had sufficiently recovered the exploration party again descended No. 1 Pit shaft. It was now almost twelve hours since the explosion. They had reached about a mile and a half inbye towards the face where the missing men were known to have been working. On reaching the stables they found that all of the ponies were dead. Exploring three quarters of a mile further forward they found that part of the Hutton Seam was on fire. The fire raging in the Hutton Seam was at a point, approximately, under the highest hill in the region at Warden Law. The inevitable conclusion was made that every miner in there would either have been killed by the explosion or would have suffocated from the afterdamp or the smoke from the fire. Mr Corbett promptly took the decision in consultation with Mr Willis, Inspector of Mines and other mining engineers to stop the progress of the fire. A party of masons volunteered and despite the danger of further explosion and exposure to afterdamp they built brick stoppings nine feet deep to seal off the oxygen feeding the fire raging on the other side of the wall.

The bodies of the victims found by the exploring party and brought to bank were: -

Thomas Hutchinson junior, 25 years old, Stoneman
Robinson Hunter, 45 years old, Shifter
Thomas Spence, 39 years old, Master Shifter
Charles Lawson, 28 years old, Furnaceman

The bodies of the following men were left in the pit behind the stoppings and there was not the slightest doubt by the explorers that they would all be dead: -

Ralph Hepplewhite, 55 years old, Shifter
Thomas Bousfield, 49 years old, Shifter
James Ashden, 41 years old, Hewer
John Bowden, 46 years old, Hewer
Edward Laing, 43 years old, Hewer

William Coates, 30 years old, Shifter
William Dunn, 57 years old, Hewer
John Richardson, 50 years old, Shifter
David Ballantine, 69 years old, Shifter
Matthew Brown, 38 years old, Shifter
Thomas Norris, 60 years old, Shifter
Thomas Proud, 58 years old, Shifter
George Barker, 16 years old, Putter
Thomas Dobson, 13 years old, Driver
Robert Straughair, 24 years old, Shifter
William Young, 67 years old, Shifter
George Shipley, 34 years old, Shifter
John Weddle, 52 years old, Shifter
William Robins, 28 years old, Shifter
Thomas Tones, 64 years old, Shifter
Edward Campbell, 30 years old, Hewer
John Hays, 60 years old, Deputy

Of the four men who were brought up alive a shifter, Robinson Hunter, died within an hour of reaching bank. Another, a stoneman Thomas Hutchinson senior who was found not far from the shaft was so severely injured that he was not expected to survive. He was badly burnt around his head and was sent to his house in Model Row where he was attended by Drs Beattie and Taylor. The cause of the explosion was unknown. Fears were expressed that it may have been carelessness by one of the men. Just four months earlier John Kearney, miner at Seaham Colliery, was fined 40 shillings, or one month's imprisonment, at Seaham Magistrates Court for contravening the Mines Inspection Act by being in possession of a pipe in No. 3 Pit. Naked candles and flames were not allowed in Seaham Colliery and the men were issued with flame safety lamps to illuminate their workplace. There was also the possibility that the ignition that caused the explosion could have been a shot fired by one of the stonemen. The probable cause became clearer at a later inquest. The inquest was formally opened the day after the explosion before Coroner Crofton Maynard in the New Seaham Inn. After the jury examined the body of Thomas Hutchinson the Coroner took no more evidence and adjourned the inquest to 16th November.

Four days after the explosion the furnace was re-lighted at the shaft bottom to restart the ventilation process. Within a week it was announced that certain sections of the mine could be worked safely and the men were invited to volunteer to resume work. All of the men had been laid off with no wages since the explosion. Faced with the prospect of a prolonged lay-off they had no option but to agree to the manager's request to return to work in the unaffected part of the mine whilst the bodies of their colleagues lay entombed for several more weeks. The speedy decision and action in walling off the inferno blazing in the Hutton Seam created deeply embittered feelings for many years between the pitmen and coal owners throughout the Northern coalfield.

A consultation of mining engineers took place on 13th November to consider whether the stopping should be opened in the Hutton Seam. The air gauges installed in pipes in the stoppings were examined and it was decided to defer the opening of the Seam for three weeks more. It was not until December 20th that the stoppings were eventually opened and the remaining 22 bodies were recovered.

Thomas Hutchinson senior who was so badly injured in the explosion that everyone thought he would not survive recovered to the astonishment of everyone who saw him on the day of the explosion. He was able to give evidence at the inquest and proved to be an invaluable witness to events in the Hutton Seam that night as he was the sole survivor of those in the field of the explosion. He was working with his son who was killed by the explosion. Both of them were stonemen and they had prepared a shot. He was adamant that at the very instant the shot was fired "the pit fired". The shot had exploded and the fire came both together. Mr Willis, Inspector of Mines, made a poignant remark to a witness at the inquest. The pit was known to have been described by some miners as the "Hell Pit" because of previous accidents and explosions. He put a question to one witness. "Do you know a place called "Hell"? Clearly, he wanted to open up a debate on this matter which was not being openly discussed. The witness did not answer. After a three-month delay involving two adjournments the jury gave a verdict that the 26 men died "accidentally from an explosion caused by an outburst of gas from the roof of No. 2 Bankhead of No. 3 Pit".

Quite by chance the lives of many more men were saved that day. The explosion happened in the Hutton Seam which was fifty fathoms above the main seam worked by Nos. 1 and 2 Pit shaft. A "staple" was sunk from the Hutton Seam to the seam below so that coals from the Hutton seam could be sent down the staple and sent to bank using No. 1 and 2 Pit shaft. At the time of the explosion two tubs full of coal were in the staple and the force of the explosion shattered it to pieces and completely closed up the staple. This cut off all communication with the seam above and fortunately for the men working in the lower seam prevented the fire-damp from extending into their workings.

CHAPTER 7

Nine years later

The daily work at Seaham Colliery

The next chapter takes the reader nine years forward from the 1871 explosion to even more horrific events at Seaham Colliery in 1880. Before we go there it might help those with no knowledge of the coal mining industry to understand the workings of the mine. This chapter will assist the reader to understand some of the working arrangements discussed in the following chapters by describing the organisation and management of the mine at that time and introducing the names of some of the individuals and their positions at the colliery.

The organisation and span of control of the senior management team in the mine was as follows: -

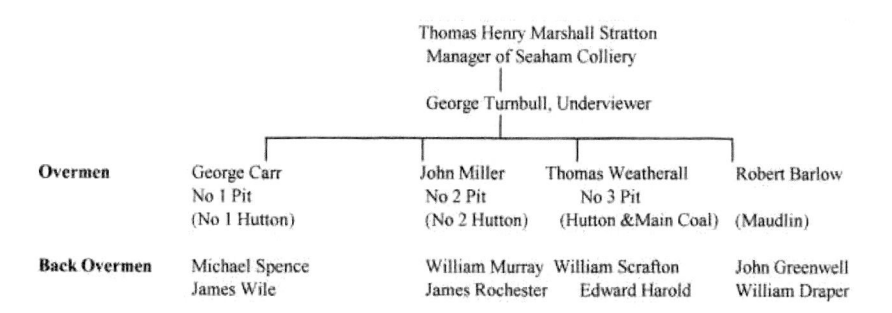

		Thomas Henry Marshall Stratton Manager of Seaham Colliery		
		George Turnbull, Underviewer		
Overmen	George Carr No 1 Pit (No 1 Hutton)	John Miller No 2 Pit (No 2 Hutton)	Thomas Weatherall No 3 Pit (Hutton &Main Coal)	Robert Barlow (Maudlin)
Back Overmen	Michael Spence James Wile	William Murray James Rochester	William Scrafton Edward Harold	John Greenwell William Draper

The Manager, twenty-eight-year-old Thomas Stratton, was responsible for the whole colliery which included all operations in Nos. 1& 2 pits and No. 3 pit and all operations at bank (on the surface). He had been in that post for six and a half years having previously worked for five years at Wearmouth Colliery, six months in Cleveland and immediately prior to Seaham Colliery he was Manager at Pease's Colliery for two years. Prior to his appointment with the Pease Coal Company he gained his management experience with Richard Heckles a well know Viewer of the Earl of Durham Collieries who carried out important experiments on lighting in coalmines and on the Davy safety lamp

Thomas Henry Marshall Stratton – Sketched by the Illustrated London News 1880

George Turnbull, Underviewer is in general charge of the mining operations and is expected to examine the workings of the mine daily. This inspection would also include the waste; the state of the air courses; ventilation and the quantity of air passing each work station. He would receive a daily report from the overmen and the master wastemen on the operations under their charge. From information provided by those reports he is responsible for correcting irregularities and obviating all discoverable sources of danger.

The four overmen assisted by the eight back overmen have the charge of the working of the pit, and more especially of the safety of the men. Their duties include coal production and development, attending to the lighting of the pit, inspecting lamps and directing whatever lamps should be used in exploring drifts or while working.

The Deputies assist the overmen. At Seaham Colliery in 1880 there were:-

No. 1 Pit – 14 Deputies
No. 2 Pit – 12 Deputies
No.3 Pit Hutton Seam – 4 Deputies
Main Coal – 4 Deputies
Maudlin Seam – 14 Deputies

Their duties are to go into the pit every morning, one hour before the hewers, to examine every working place in the pit and in particular to ascertain that it is in a safe working condition. They also examine all safety-lamps before passing the "caution board," and lock them so that the men cannot tamper with them. No hewer is allowed to enter his working place until he has been examined by the deputy of his district. The deputies also inspect the laying of the tramways and the securing of the workings by timber. At the close of each day's shift the deputies see all the men and boys out of their respective working-places, check that no lights are left in the pit, that the ventilation doors are closed and that the ventilation is in good order.

A number of other officials work underground. The other officials in 1880 were: -

Master Wasteman – Joseph Spence

The Master Wasteman travels daily around the old workings of the pit to see that the air-courses are in good order. His main role is to remedy any faults in the roof and generally attend to the ventilation systems. Two assistant wastemen traveling together would be expected to go over the whole of the waste in the pit at least once a week.

Four Master Shifters – Charles Dawson No. 1 Pit (No. 1 Hutton Seam); William Hartley No. 2 Pit (No. 2 Hutton Seam); Anthony Smith No. 3 Pit (Hutton Seam Main Coal); Walter Murray (Maudlin Seam)

The four Master Shifters are in charge of the Shifters who repair the horseways and other passages in the mine keeping them free from obstructions.

There were three coal-loading shifts at Seaham Colliery in 1880. They were: -

Fore Shift – 4.00 am to 11.30 am

Back Shift – 10.00 am to 5.30 pm

Night Shift – 4.00 pm to 11.30 pm

The Master Shifter and the Shifters work in four teams one for each district. They are the "repairing shift" and do not work regular coal-loading shift times as above. Instead, they descend the mine at 10.00 pm and finish their shift at 6.00 am.

The Engineer, Wilkinson Rowell, is in charge of two enginemen in No. 1 pit, one engineman in No. 2 pit, the brakesman, engine wright and the pumping engineman and the operation of the furnaces and boilers below ground.

The output of Seaham Colliery in 1880 was 500,000 tons of coal per year or, on average, 2,500 tons per day. The colliery had a total workforce of about 1,500 men and boys of which 480 were hewers working on the coal face winning the coal. The area of the workings underground was quite significant. The main coal seam covered 480 acres; the middle level worked in the Hutton and Maudlin seam in which the explosion occurred covered 500 acres and the lowest level, the Harvey or Hutton No 2 seam covered 700 acres.

The ventilation in the mine was achieved through the two shafts. The shaft for No.1 and 2 pits was the downcast shaft. This downcast shaft was divided by a brattice. One half of the shaft served the highest and the middle level and the other half served the lowest or No. 2 Hutton level. The return air came back up from No. 3 pit i.e. the upcast shaft. The movement of air throughout the mine was achieved by two furnaces which were about 150 yards apart both on the level of the Hutton seam. The No. 3 pit furnace was situated close to No. 3 pit shaft and the No.1 pit furnace was situated close to No. 1 pit shaft. These furnaces were about nine feet by seven feet and they each burned about ten tons of coal daily. The arrangement of the furnaces and ventilation system of the mine is shown on the sketch below.

The air flow created by the furnaces was boosted by several boilers and in total the furnaces and boiler fires achieved an air flow of between 320,000 to 330,000 cubic feet per minute. The boilers were usually worked between fourteen and twenty-one hours per day and the smoke from the boilers went up No. 1 pit furnace staple. In addition to the boilers there were two semi-portable engines; one called the Maudlin engine about 100 yards from No. 3 pit shaft and the other is called the No. 3 Hutton seam engine and was situated close by No. 3 pit shaft. Each day one of the furnaces was damped down and cleaned out. It took about one hour to complete the damping down, cleaning out and re-starting the furnaces. During this time the air flow would be reduced but the Manager, Mr Stratton, was confident that the ventilation would not be materially affected during the downtime.

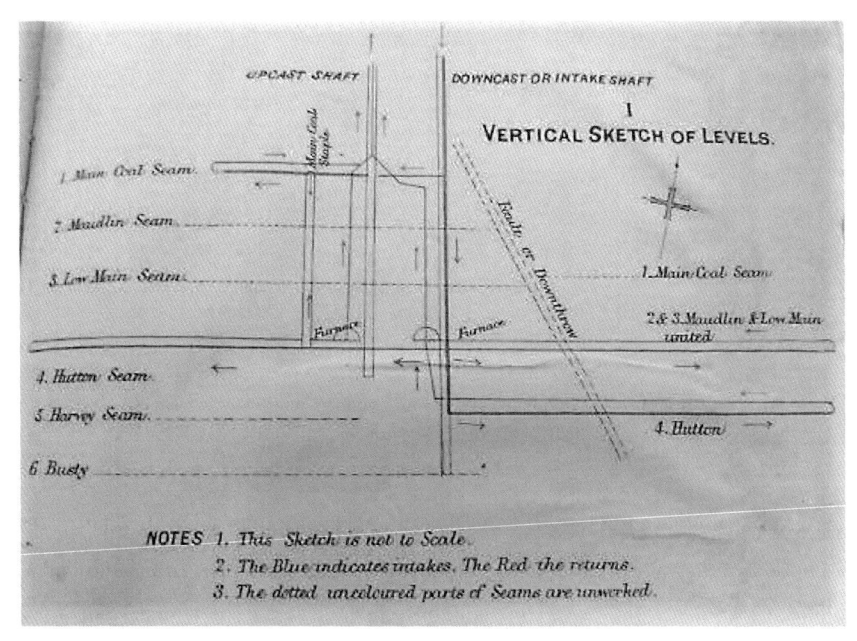

Sketch of ventilation system in the mine
Courtesy of Durham Record Office - DRO D/X 1051

The ventilation of coal mines by furnace-fires was the accepted method of working in the collieries of County Durham. The 1911 Coal Mines Act prohibited the use of furnace fires and from that date mechanical ventilation using fans was adopted.

Barometers were fixed at the top and bottom of the pit shaft and daily readings were taken. In bad weather a sudden fall of the barometer was immediately reported to the Overmen in charge of each district. Once alerted they would keep a strict watch in case fire-damp should appear owing to the light state of the atmosphere. Explosions of fire-damp have invariably occurred during periods of significant atmospheric change. If, after testing, the Overman or Deputy knew where gas was accumulating, they would dilute the concentration of gas by the addition of a greater volume of fresh air through the workings. In the very early part of the 19th century some coal owners paid a "fireman" to disperse any gas found by igniting it with a long stick with a lit candle on the end. They literally risked their lives and those of their colleagues every time they entered a gaseous part of the mine with their lit candle.

The explosion in 1871 killing 26 men and boys was a turning point in the use of candles and naked flames at Seaham Colliery. Strict discipline and attention to standing orders was observed in the use of lighting. It had been clear that throughout the coal industry the lives of thousands of pitmen had been sacrificed by the careless use of open flame lights. Despite a code of rules that were established to protect every man and boy in the pit the majority of men were sometimes put at risk by the careless few who would selfishly indulge in their own pleasures by taking pipes and tobacco into the pit. William Crawford, the Secretary of the Durham Miners Association in the July 1881 monthly circular to the miners of Durham refers to an instance of gross and culpable carelessness which occurred at Ryhope Colliery when a box of matches came to bank on top of a tub of coal. He declared that "no punishment is too severe for a rascal who is so flagrantly guilty of such a gross violation of the law and where the lives of hundreds of other persons are endangered". The Ryhope men took immediate action by offering £1 reward to any person who would give such information leading to the detection of the culprit. In 1882 Patrick Farrell was fined twenty shillings plus costs for having a tobacco pipe in his possession in the pit. Some years later the miners at Seaham Colliery were outraged when a box of matches was discovered down the mine. The men and the colliery management offered £10 to find the miscreant who took the box of matches down the mine

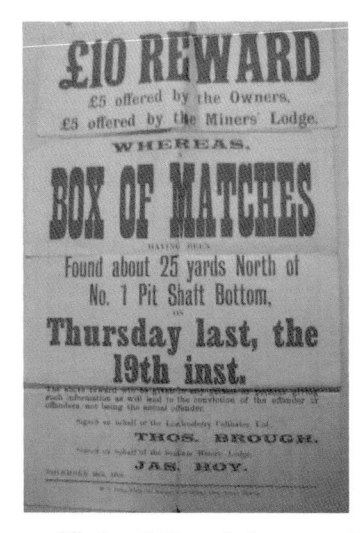

Notice posted around Seaham Colliery offering a reward for information

Over many decades numerous versions of flame safety lamps had been designed and proposed to cast a safe light in coalmines. The most commonly used lamps were the Clanny; the Davy and the Stephenson flame safety lamp. The main principle in all three was that gas could be tested for by raising the lamp up high where the lighter methane gases would accumulate and observe the flame. If gas was present the flame became elongated and burned blue. Dr William Reid Clanny developed his safety lamp whilst working at Sunderland Infirmary. It was rumoured that Sir Humphrey Davy produced his version of the safety lamp shortly after visiting Dr Clanny in 1815. Dr Clanny was the visiting physician at Seaham Infirmary from 1844 until his death in 1850. Through experimentation Sir Humphry Davy adapted his first lamp when he realised that a flame could not be passed through wire gauze containing from six to eight hundred holes per square inch. At Seaham Colliery the Clanny lamp was predominantly used by the hewers and others working on the coal face whilst shotfirers would carry the Davy lamp. It was generally believed that the light from the Clanny lamp was better that the Davy lamp. A later improved version of the Davy lamp called the "Jack" Davy lamp which gave off a slightly better light was used by the Deputies. It was essentially the same as a Davy but it had a glass shield instead of a wire shield. In 1880 the colliery had 900 Clanny, 450 Davy and 150 Jack lamps. Every hewer had his lamp inspected by the Deputy every morning before he went into the pit. In the case of the other men the overman inspected the lamps.

The Stephenson, Davy and Clanny lamp

The examination procedure required that every man took his lamp to pieces. Every part was taken to the Deputy who inspected the gauze and the two leather washers. If he found them in good condition, he returned them to the men who put them back together again and offered them back for final inspection. The wick in the bottom section was lit and screwed on to the top. The Deputy then locked the lamp so that it could not be opened underground.

At Seaham Colliery in 1880 naked lights were allowed in some places. The horse keepers carried naked light glass lanterns in the area outside the caution boards which in the case of the No. 1 pit was about 400 yards from the far-off end just 30 yards from the far-off furnace. Other caution boards were placed at No. 3 East end and also in No. 3 Hutton. Naked lights were also allowed in the north Maudlin until about 200 yards from the end including at the Polka turn. All of the stables except the in-bye Maudlin stables had naked flame glass lamps.

Horse keeper working with a naked lamp (Fordyce, 1860)

Ironically on 15th October 1880 a miner, William Cowell, pleaded guilty at Houghton-le-Spring Magistrates Court to a charge of having a naked light 600 yards beyond the caution board in the Hetton pit. So seriously was this conduct viewed by the pit management that the magistrate sent Cowell to Durham Gaol for one month with hard labour.

Although the jury in the 1871 Seaham Colliery disaster gave a verdict that the 26 men died "accidentally from an explosion caused by an outburst of gas from the roof of No. 2 Bankhead of No. 3 Pit" it prompted much debate about the use of "blasting" and a number of changes were introduced to shotfiring procedures. The use of blasting was permitted in stone but no shots were allowed to be fired in coal. In many instances the men could not work through stone without blasting. If a stoneman was required to excavate or work through stone then the Deputy or Overman made an application to the Underviewer who consulted the Manager. If the Manager considered it necessary and safe to fire shots then he would give a written permission; a named person was selected to fire the shots and he was given instruction when and what to do and how to do it. After the special rules appertaining to shotfiring were drawn up they were reviewed and approved by Mr Bell, the Inspector of Mines. A copy was then given to each shotfirer. Shots were fired every day in many different parts of the colliery particularly to widen stone drifts and to make refuge holes. The rules did not require a shotfirer to damp that part of the mine where a shot was to be taken. When giving permission for shots to be used the Manager took account of a general rule in the Coal Mines Regulation Act which stipulated that shots should only be made in places where "ventilation is so managed that the return air from the place where the powder is used passes into the main return air course without passing any place in actual course of working".

The general management and day-to-day operation of the mine in 1880 has been described in the above paragraphs. The reader with no previous coalmining knowledge should now be better able to put the tragic events that are set out in the following chapters into context. The 1880 Seaham Colliery Explosion ripped the heart out of the colliery village and changed the lives of so many families forever.

CHAPTER 8

The beginning of one year of hell
Day 1,   8th September 1880

The colliery had enjoyed nine years of immunity from any serious explosion since the twenty-six men and boys perished in the October 1871 disaster. It was 2.20 am on the morning of Wednesday 8th September 1880 when an explosion was heard in the colliery village and the ground vibrated as though there was an earthquake. The violence of the explosion was heard by sailors on ships in the harbour; by miners at Murton Colliery and for a great distance around the district. Almost everyone in the village was awakened and within a short period of time large crowds began to gather around the mouths of the two pit shafts desperately trying to find out what had caused so much devastation around the pithead.

Anxious crowds gather around the pithead (The Illustrated London News)

The next shift of men was due to go down at 4.30 am but, aroused by the blast, they were on the scene long before their start time. Clouds of soot and debris had been thrown out of the shafts and prevented any access to the head of the shaft. Thomas Stratton, the Colliery Manager, was immediately sent for and woken out of his bed. On arriving at the colliery, he organised the pitmen who had arrived early for their shift to clear away the debris to get to the shafts. On inspection of the downcast shaft it was found to be badly wrecked with both No. 1 pit and No. 2 pit cages fast with the guide ropes and the rapper ropes badly damaged. Mr Stratton instructed the engineman to bring the cage of No. 3 pit to bank but it was found that the guide mechanism within which the cage sat was completely broken. At this point the grave events that happened at Hartley Colliery in 1862 in which the shaft was blocked leading to the suffocation of all the men in the mine must have entered the thoughts of Thomas Stratton. The enormity of the disaster was beginning to emerge. There were 231 men and boys down the mine.

Mr Corbett the Principal Viewer was sent for and messages were sent to other mining engineers and colliery viewers in the district to give assistance. On returning to No. 2 pit shaft the Manager sought out a jack engine which was quickly fitted to raise and lower men into the shaft on a rope. It was now 6.00 am and the wives and families from almost every house in the village were now at the pithead anxiously waiting for news. Thomas Stratton and a shaftsman called Copeland were lowered down No.2 pit by the jack engine on a rope with knotted loops. They inspected the shaft for damage descending half way down between the surface and the first level. Returning to bank another two shaftsmen joined them to explore further down. This time they reached the main coal level. On returning to bank, they were able to give the good news that they had communicated with men who were working on that level and who were waiting for rescue close to the shaft. The information was soon passed to the anxious waiting crowds that there were about seventy men and boys from the main coal level waiting to be brought up the shaft. All of them were uninjured with the exception of one man, William Laverick, who was badly burnt. However, Mr Stratton and the other shaftsmen had explored further down beyond that level. They found the shaft completely and utterly blocked by tons of the shaft lining from above and there was absolutely no access between the main coal level and the Hutton level. They had shouted out but had received no response from the estimated 180 men believed to have been working in the Hutton level.

The men at bank had not been idle. A kibble (an iron bucket) was fitted to the jack engine and gangs of shaftsmen descended the shaft almost continuously until eventually 19 relieved and exhausted survivors from the main coal seam were brought up the Low Pit Shaft.

The nineteen survivors brought safely out of the Main Coal Level were:-

Andrews, George	Northway, James
Bell, Richard	Smith, Edward
Bell, Richard	Smith, Thomas
Cowley, William	Strawbridge, Robert
Johnson, Thomas	Surtees, Edward
Johnston, William	Thompson, George
Kirkbright, William	Wardle, Robert
Mann, David	Wilkinson, Thomas
Marley, Ralph	Wilson, Charles
McKay, J	

Meanwhile in the High Pit Shaft – the No. 3 pit - another kibble was fitted and the shaftsmen diligently descended time and again until another 48 men from the Harvey level were safely brought out alive. They had made their way up from the Harvey by means of a stone drift to the Main Coal level. One of those, William Laverick, an onsetter had suffered terrible injuries from the explosion. His face and head were swollen to an enormous size so much so that his eyes were only just visible and all of the hair on his head and face had been scorched off.

Underground access route between levels at Seaham Colliery

The forty-eight men and boys brought safely out of the Harvey Level were: -

Brown, George	Lamb, Henry
Brown, George	Laverick, William
Cairns, J	Mason, John
Chapman, Matthew	Miller, Henry
Crane, Edgar	Morris, William
Cummings, William	Muncaster, Matthew
Curry, Robert	Osborn, Robert
Dillon, P	Pellew, Henry
Dixon, Thomas	Proctor, Robert
Dodds, Thomas	Quayle, Joseph
Forster, Mark	Riley, William
Forsyth, Samuel	Steel, Jacob
Gardener,	Stephenson, John
Gatenby, John	Taylor, Joseph
Graham, John	Telford, William
Greener, T	Turnbull, John
Hall, John	Vickers, Thomas
Hartley, William	Wilson, Robert
Henry, Thomas	Wilson, William
Horsfield, T	Wilson, W
Howe, Stephen	Winter, William
Hunter, William	Wood, George
Johnson, Thomas	Young, George
Kent, Alexander	Young, Robert

By 9.00 pm that evening a total of 67 men had been rescued from the mine. The crowds of people who were gathered around the pithead had quietly exhibited every sort of emotion that a human could endure. Those whose husbands and sons had been rescued from the main coal seam were joyous and thankful but for others whose men were working in the Hutton and Maudlin Seam there were tears in their eyes and unmistakable signs of terrible inward suffering, frantic grief and fear for the fate of their loved ones.

There was a plentiful supply of experienced and practical shaftsmen to hand. Fresh teams were waiting at the pithead to relieve those exhausted shaftsmen returning to bank. All of the neighbouring collieries had swiftly sent their shaftsmen to the scene to lend assistance in addition to quite a number of colliery managers and mining engineers.

Teams of shaftsmen descend to unblock the shafts
(The Illustrated London News)

Mr Corbett, head viewer, supervised the arrangements at No. 3 pit shaft whilst Mr Stratton took charge of No. 1 and No. 2 pit shaft. By the end of that morning the Marquess of Londonderry, Mr Forster (South Hetton Colliery), Mr Hall (Haswell Colliery), Mr Bailes and Mr Gambier (Murton Colliery), Mr Atkinson (Assistant Inspector of Mines), Doctors Broadbent of Murton and Beattie from Seaham and many other mining engineers were available to provide whatever assistance was necessary.

It was now clear to the rescuers that the explosion did not happen in the highest level, the Main Coal or in the lowest level, the Harvey Seam as the men working there had survived and were rescued. The scene of the catastrophe was in the middle level either the Hutton or Maudlin or both. Survivors who came to bank declared that they had seen many dead bodies and some in a state that showed the force of the explosion had slain them. Thomas Stratton, the Manager declared "Time is everything if the men are imprisoned alive. God help the poor fellows down below, and God help the anxious awe-struck relatives and friend above". He knew how remote the chances of finding the other miners alive were. First the blast from the firedamp passing along the level would slay everything in its path. Secondly there was the risk of the workings collapsing and the roof falling in and crushing the men. Thirdly, the ventilation down the shaft had been cut off for many hours and there was every likelihood that afterdamp had miserably smothered the victims. As the shades of evening fell on that first day there was satisfaction and relief that some men had been brought to bank from the Main Coal and the Harvey level but great anxiety was felt by the fact that as yet no communication had been made with men in the Hutton and Maudlin seam.

While efforts were made by relays of men frantically working to clear the blockage in both shafts an exploring party had made its way into the Main Coal

Seam and slowly progressed up to the Main Coal Staple. Here they found that the stables were on fire and they knew beyond doubt what would be the fate of those poor horses in the stables. Meanwhile progress was hindered in the No. 3 pit shaft when it was discovered that below ground the engine house was on fire preventing access into the Hutton and Maudlin Seam. Teams of firemen, volunteer miners, were sent down to fight and extinguish the fires.

The President, Secretary and Treasurer of the Durham Miners Association, Messrs W Crozier, WH Patterson and J Foreman, arrived at the colliery and arranged for two of the local lodge officials to act on behalf of the men and to accompany each of the exploring parties. Mr Patterson and Thomas Burt went with the first shift; Mr Crozier and Mr Foreman with the second and Mr James Wilson and Thomas Banks with the third. Thomas Burt and Thomas Banks, lodge officials, were to play a prominent part in the events at Seaham Colliery over the next year. (See photograph of these five explorers overleaf)

The crowds on bank shuddered as hastily made coffins were carried by the colliery joiners to the pithead for those who were not expected to survive. The crowds around the village steadily increased that afternoon.

Trains arrived from all over the colliery districts bringing people expecting to attend the Seaham Flower Show that day. It had been cancelled because of the explosion and instead those people made their way to Seaham Colliery through morbid curiosity adding to the dense crowds mingling around the village.

Many miners including quite a number of the men who were also members of the 2nd Durham (Seaham) Artillery Volunteers had hoped to attend the flower show and had swapped shifts with other miners. This resulted in confusion about who was actually in the pit at the time of the explosion. Fate had intervened putting them at work down the pit when the explosion happened instead of being fast asleep at home in bed.

Thomas Burt, Thomas Banks and William Crozier of the Seaham Lodge with Messrs Patterson and Foreman of the Durham Miners Association - Courtesy of Seaham Family History Group

CHAPTER 9

The loss of life is considerably larger than first thought

Day 2 to 3,   9th to 10th September 1880

Amid the confusion and chaos on the day of the explosion unofficial estimates of men still missing and thought perished was thought to be 120 but many at the pit were uneasy that a considerably higher number might have been sacrificed. These fears proved well founded. Throughout the afternoon and evening colliery officials visited every house in the colliery village to ascertain the names of men and boys missing from their homes. By late evening it was clear that the number still missing was nearer 160 than 120 and that those families in addition to the grief and heartache that was to come would also experience severe financial hardship. Quite appropriately and timely representatives from the Board of Management of the Miners Permanent Relief Fund arrived at the colliery. They were permitted to take up quarters in one of the workshops so they could be on hand to administer help and relief to any families in need and they accompanied colliery officials visiting the homes of missing pitmen to ascertain the dependant relatives and the consequential liability on the fund.

Work continued all through the day and the night to clear away the blockage at the high colliery No.3 pit shaft. At the twin shaft of the low colliery No. 1 and 2 pits it became clear that the shaft was terribly shattered just above the Main Coal Seam and would take weeks to repair. Working parties ascended and descended the shafts in monotonous succession returning exhausted and drenched with water. At the No. 3 pit the work was entirely connected with clearing away the debris from the workings at the bottom of the shaft and restoring the ventilation. Reports were constantly reaching bank that the fires raging below had been extinguished and then shortly afterwards they were burning again.

On Friday, Day 3, Mr Grounds a viewer in charge of an exploration team returned to say that they had broken through to the foot of the shaft and found several bodies in all stages of mutilation including a severed head and leg that could not be connected to other bodies at that spot. One of the miners in the rescue team, Mr Jobson, told reporters that his party went a hundred yards along the Maudlin Seam to the engine house and came across bodies severely disfigured. Two or three of the team had been charged with wrapping up the bodies in canvas and carrying them to the shaft bottom with a ticket attached with the names if they could be identified. In some cases, it was only possible to identify a corpse by the metal tally on their belt or lamp. When they were properly wrapped, they were put into the kibble in two's and sent to bank.

Sketch of the first bodies brought to bank (The Illustrated London News)

The kibble rose to the shaft head carrying two bodies placed upon the edge of the tub in the fashion of living men. The tub was brought to bank amid oppressive silence and then solemnly and quietly, as though in fear of disturbing their rest the miners gently and reverently transferred the remains of their dead comrades onto stretchers and carried them away to the shed where the coffins were stored. Then followed a pause of about an hour and a repetition of the same scene took place.

Returning from the Maudlin engine house Jobson and the exploring party went up the way to No. 3 Pit Staple where they found the body of young William Venner and knowing that he had been working with his father they looked around but the explosion had blasted all before it and the search was fruitless. They found the trousers and other clothing of a boy called Lawson and it was suspected his body would be under the chaos and debris with others including Samuel Venner, Robert Potter and Thomas Lowdy the night shift Deputy. Pushing further forward the body of Anthony Smith was found at the high pit staple together with the body of Joseph Rollins. The body of Walter Dawson was found at the high pit furnace and shortly after the bodies of Joseph Straughan and John Mason were found in No. 3 Hutton and Thomas Forster was found in No. 1 Hutton. There were altogether fifteen to twenty bodies in close proximity to the shaft. Amongst them were John McGuiness a hewer; John Hunter, furnaceman; a hewer called Lees Ball Dixon who had made a pillow of his coat and laid his head down on it and Thomas Gibson a fireman who left a large grieving family. Gibson's body was found in a fire hole covered by debris and his legs were so crushed the rescuers had to get a special coffin to bring it to bank. Both John Hunter and Thomas Gibson had only worked at the pit for a short time having previously left their jobs at the Seaham Bottleworks because of a lull in the trade. The horse-keeper John Neasham was found in No. 3 Hutton stables. The bodies of James Kent and William Wilkinson both putters;

Thomas Williams a fourteen-year-old driver; Thomas Lowdey, deputy and then Richard George who left a widow and five children were discovered. Thomas Lowdey had been caught by the explosion under the chin and his head was almost taken off. The body of Richard George was sent to bank in the kibble accompanied by two parcels, one containing two arms and the other containing one arm. The attendants at the dead house had been instructed to strip the canvas from the bodies as they arrived and to wash the corpses before sending them in coffins to the homes of their family. It was found impossible to carry out this instruction. When the canvas was removed from the bodies it revealed a pitiful sight. The corpses were burnt from head to foot; the limbs were shrivelled up to half their normal size or in those cases where the burning had stopped at a certain stage were swollen by the intense heat. When the attendant attempted to wash the dirt off it was found to have been scorched into the skin and nothing would remove the ebony-like appearance. Liquid disinfectant was poured into the coffins and over the bodies.

The survivors' spoke of a man called George Dixon who refused to take the opportunity to escape to safety with the rest of his workmates. He was uninjured but had refused to leave his workplace with his marra's because his putter boy Robson Lawson had been injured and could not be moved. The exploring party found Dixon with his arm around the boy both dead from the effects of afterdamp. He had nobly sacrificed his own life to stand by and protect his putter boy. By the end of that Friday thirty-one bodies had been recovered.

Behind the drawn blinds of many houses in the colliery village there were heartbreaking tales. In many streets and rows there were as many as six or seven pitmen known to be either dead or still missing and presumed dead. One poor victim, a chock drawer named John Sutherland living in Australia Street had left a grief-stricken widow and nine children; one poor woman whose husband was amongst the missing went into confinement on the same day the explosion occurred; in one house a father, five sons and two lodgers were taken and one unlucky fellow called Watson, a shifter, had only worked eight shifts at the pit. Thomas Hutchinson, shifter, of Seaham Street survived the explosion of 1871 in which his son died but fate took its revenge and caught up with Thomas. Other stories tell of men on whom divine providence smiled that day. The choirmaster and organist of Christchurch, Joseph Birkbeck, was due on shift that fateful night. He slept through the caller and missed his shift. Later in life he was to become Postmaster at Seaham Colliery living to the fine old age of 93 in Birkbeck Villa. Evidence brought to the attention of the jury at the later inquest revealed that one survivor brought out of the Main Coal level, Thomas Johnson, had survived four earlier explosions at the pit. Illness was to be the saviour of another survivor. A hewer in the Maudlin Seam, John Hutchinson, was ill when he left his house to start his shift at 10.00 pm. Faced with the prospect of losing money or working through his illness he took his lamp and went down in the cage. Progressively his illness made him feel worse until eventually he decided to make his way outbye. The overman, Walter Murray, came across him at 2.00 am half asleep along the travelling road and ordered him to catch the next cage and get home to his bed. Eventually John reached the shaft bottom and chatted to the Onsetter Laverick before getting into the cage. Just two minutes after he stepped out at bank the pit fired.

Anyone waiting at the shaft top that day would have seen the sad sights of ghastly corpses being hauled up from the pitch blackness. All of the bodies returned to bank that day were removed to the temporary "dead-house" at the colliery. This was a low whitewashed building normally used for stores.

Wives and families waiting to identify bodies at the "Dead House"
(The Illustrated London News)

After identification the bodies were sent to the grieving family homes. Onlookers watched the sickening sight of carts loaded with hastily made coffins pushed slowly through the village. A message of sympathy sent from Queen Victoria at Balmoral Castle to the Marquess of Londonderry was posted at the pit-heap. The Queen asked if any more lives had been saved and what was the cause of the terrible explosion. She asked the Marquess to "convey to the relations of the missing men her sincere sympathy in their distress".

Mr Corbett, the colliery viewer worked untiringly in organising men and materials needed for the repair and exploration operations. He was most anxious to join the rescue teams underground but was persuaded by the management committee that his services were best used at bank. Meanwhile matters below ground were becoming serious and a consultation of all the mining experts present discussed the situation of the foulness of the air. It was becoming dangerous for the men to push their way further into the workings. The only way the exploration could continue and further bodies could be found was by improving the ventilation and the agreed advice was to stop the exploration into No. 3 shaft for the night and to resume the next day.

CHAPTER 10

The first inquest is opened

Day 3, 10th September 1880

Thirty-one bodies had been found and brought out of the pit. The wives and families of the victims needed to bury their dead. However, the laws of the country require an inquest to be heard for any unexplained or sudden death. In order that the burial of the dead could go ahead without undue delay arrangements had to be made to carry out an inquest as soon possible.

Crofton Maynard who presided over the 1871 Seaham Colliery Disaster inquest was working on an inquest at Haswell Colliery. Mr Bell, HM Inspector of Mines telegraphed him to ask if he would head the inquest into the death of the bodies recovered. He came at once arriving at 6.30 pm at the New Seaham Inn on Londonderry Road (now Station Road) just over 100 yards from the colliery entrance. He immediately opened the inquest on the bodies of James Brown, stoneman of Hall Street and Thomas Forster, shifter of Doctor's Row and it was anticipated these would be the test cases on which all of the other deaths would be judged. John Richardson was appointed foreman of the jury and other jury members empanelled were: - Thomas Elliott, George Stokeld, Thomas Chilton, Cuthbert Watson, Samuel Cockburn, Francis Marshall, David G Smith, John Softley, Thomas Forster, James Clare, George S Wallace, Johnson R Thompson, George Grieves, Robert Graham and James Ayre.

Representing the interested parties at the inquest were: -

Mr HB Wright (Solicitor), Mr JR Eminson (Chief Agent), Mr Corbett (Head viewer) on behalf of the Coal owner the Marquess of Londonderry.

Mr Thomas Bell (Inspector) and Mr A Atkinson (Assistant Inspector) representing HM Inspectorate of Mines

Mr Pickard representing the Miners Permanent Relief Fund

In his opening address to the jury Crofton Maynard, Coroner, set out the object of the inquest into the frightful catastrophe. It was to ascertain the cause of the disaster and he intended that it would be as complete and thorough as possible. That night he proposed that the jury should view the bodies of James Brown and Thomas Forster and take evidence to confirm their identity. As for the rest of the bodies brought out of the pit, he declared that as long as meticulous records were kept of the identification of any recovered bodies, he was happy that the families could proceed with interment without any formal viewing by the inquest jury members particularly in view of the decomposed and mutilated condition of some of the victims.

After seeing and hearing identification of Brown and Forster the Coroner gave the formal order of burial. He then set out his approach to the inquest. He would hear evidence from colliery officials and also representatives of the men. It was important that evidence was heard as soon as possible from the men brought out of the pit alive so that matters were fresh in their minds. Mr Wright, Solicitor for the owners pointed out that the colliery officials were engaged in rescue and exploration work and could not present themselves to the inquest for some time. In consultation with the jury the Coroner agreed to adjourn the inquest until Wednesday 15th September and for it to be reconvened at the Londonderry Literary Institute.

CHAPTER 11

The burials begin

Day 4, 11th September 1880

Good news was brought to bank from the explorers in the Main Coal seam. Some of the ponies and horses in the stables were found to be alive and in good condition. There were fifty-four ponies alive in Number 2 Hutton and the Main Coal seam. Food and water was taken down for them and they showed evident signs of pleasure when the exploring party got to them.

Relays of men descended and ascended No. 3 pit shaft but for most of them all they could do when they reached the bottom was to watch the course of events unfolding as the teams pushed slowly into the Maudlin seam. The gas-lights at bank were put out for fear of an accumulation of gas making its way up the shaft and igniting. Through 600 yards of heavy falls of stone and timber the explorers pushed into the Maudlin seam. They came across the body of William Breeze, horse keeper, lying on his back in the Maudlin stables.

Of the bodies already recovered all but two unidentified corpses remained in the "dead-house" the others being removed to their homes. Neither of the unidentified corpses had a scrap of clothing left; both were badly burnt; one had no head and the skull of the other was torn off leaving only the jaws fixed on to the neck. A notice posted outside of the temporary office of the Miners Permanent Relief Fund stated that burial money would be paid on application by the relatives of those bodies recovered. In Christchurch graveyard a huge gloomy trench had been dug for the victims close by the monument erected to the twenty-six men killed in the 1871 explosion. On that monument was the following text from the New Testament Mark xiii, 33 "Take ye heed, watch and pray, for ye know not the hour". This was so sadly appropriate.

Concerns about the disaster in the Houses of Parliament prompted Sir William Harcourt, the Home Secretary to visit the colliery. After viewing plans of the mine and accompanied by Mr Bell, HM Inspector of Mines, and Mr Corbett, Agent for Lord Londonderry they walked to the upcast shaft No. 3 pit to view operations as exploration work had been suspended in No. 1 and 2 pit shaft. There they saw a body brought up by the manager Mr Stratton. The Home Secretary had a few minutes conversation with Mr Stratton on what was being done below in the mine. Mr Bell commented to Sir William Harcourt that he had considerable experience in coal mining in Lancashire and other parts of the country and in his opinion the collieries in Durham were managed best of them all. He did observe, however, that he considered Seaham Colliery to be one of the more dangerous collieries in the district because of the large quantity of gas to be found in the workings.

Across the road from the colliery entrance was Christchurch the parish church for the colliery district. Here the bodies of James Brown and William Simpson were interred on Saturday 11th September and a further twenty-three bodies were buried on the Sunday. At St John's Church the burial of George Dixon, Thomas Gibson, John Hunter and James Kent took place on Sunday 12th September whilst Robson Lawson was buried at South Hetton. Excluding the two unidentified corpses in the dead house these were the whole of the bodies recovered up to that date. At Christchurch all of the interred were buried in the communal grave with the exception of eleven who were buried in private graves. The eleven buried in private graves were Lees Ball Dixon, John Mason, Joseph and Robert Rawling, Anthony Smith, John Thomas Patterson, Anthony

Ramshaw, Samuel Venner, William Venner, Joseph Straughan and Richard George.

The inside of Christchurch was draped in black and as the bodies were brought in the organist Joseph Birkbeck, the man who cheated death by sleeping through his caller, played the "Dead March in Saul". The Rev. WA Scott, Vicar of Christchurch officiated throughout the ceremony assisted by the Rev. WD Allen of Dalton-Le-Dale; the Rev. GA Ormsby of Rainton; the Rev. H Martin of St John's, Sunderland; the Rev. WE Scott of Hawthorn; the Rev. H Lunn of Bishopwearmouth and the Rev. H Jones, Curate of Christchurch. The choir which had lost Robert Potter, John Lock and George Norris in the explosion sang "Thy will be done", "The Church has waited long", "When our heads are bowed with woe", Jesu, lover of my soul" and many others.

The Rev. WA Scott told the crowded church that he had lost the flower of his congregation in the disaster but he impressed on the mourners that there were no religious difficulties in the arrangements made for burial at the Seaham Colliery parish church. There were many non-conformists in the pit village many of whom were staunch Methodists. The first meetings of the Christian Lay Church had been held in miner's cottages and homes in California and Henry Street during the previous three years. The church, later built at the bottom of Seaham Street as The New Seaham Independent Methodist Church, lost many members because of the explosion. William Thompson of Murton, one of the men involved in the sinking operations had begun a "Bible Christian" society whose members met in a miner's cottage in William Street. Later, as the membership increased, they held their meetings in the canvas and rope store at the colliery until they built the Jubilee Primitive Methodist Church at the bottom of Seaham Street. The primitive Methodist preachers in that chapel were known as "Ranters" because of their enthusiastic and animated preaching at the pulpit.

James Hender, a Wesleyan Methodist preacher from Barrow-in-Furness had visited Seaham Colliery in 1874 and preached at the New Seaham Wesleyan Chapel built at the bottom of Cornish Street in 1868. He wrote about his visit in the Bible Christian Magazine that "the natural product of the district is black but the moral state of the people is already white unto harvest." The Vicar of Christchurch, Rev. WA Scott, was sensitive to the religious beliefs of all of the non-conformist victims of the disaster. "There" he said, pointing to the corner "will be buried the body of James Brown, a staunch non-conformist, who has never entered my church, but who I most sincerely respect as a true and Christian man. Next to him will be my right-hand supporter, poor Robert Potter. Beside these will be Churchmen and Dissenters buried, as they lived, together."

Outside the church the road was completely blocked on either side of the graveyard by people who had come to witness the burials. There were reported to be not less than 50,000 visitors who had travelled to Seaham on foot, by brake or by train. Some could not see but were content just to be there to pay their respects to the victims of the explosion and their families. A special evening service of prayer and hymns for wives and families was conducted.

Sketch of evening service for the families of victims: The Illustrated Evening News

CHAPTER 12

The winner of the gold cup is buried

Day 5 to 7, 12th to 14th September 1880

Only one further body was recovered from No. 3 pit over the next couple of days. Anthony Ramshaw was discovered about 500 yards from the shaft in the Polka Way with his lamp shattered to pieces close to his body. Most of the other bodies were expected to be found one or two miles inbye from the downcast shaft. However, the gas in No. 3 pit continued to be a problem and it was thought impossible to explore further inbye to recover bodies until the ventilation was restored to a normal state. The work in the No. 1 and 2 pit shaft had taken the explorers down to the rubbish lying at the bottom of the shaft. Once those obstructions were removed it was hoped that the shaftsmen would be able to draw the cage away from its embedded position.

In the dead house the two remaining corpses were finally identified although there seemed to be nothing specific to distinguish the mutilated remains. One body that of Thomas Alexander was identified by the shoes he was wearing and the other was identified as Joseph Chapman although the relatives only had an instinctive feeling that it was him.

A meeting of the men was held on the pit heap at which Lord Castlereagh was present. A resolution was passed to open a subscription list for donations to relieve the immediate necessities of the bereaved. The Rev. WA Scott, Vicar of New Seaham, had also received a letter from the Bishop of Durham to ask if there was to be a public appeal for the dependants of the deceased.

At 7.45 am on Monday morning 13th September the body of Thomas Hindson was found by John Noble at the return wheel in the Maudlin seam. The body which was brought up the No. 3 pit shaft was badly mutilated with legs, an arm and the head from the jaw upwards blown off. Identification was nigh impossible and at first the coffin was sent to the home of Samuel Venner for confirmation before it was taken to Hindson's home in John Street, Seaham Harbour. His wife refused to believe the body was that of her husband and it was only after assurance by a medical man and by the adjutant of the Artillery Volunteers that she was convinced. He was only identified by his very distinctive imperial beard. Hindson, a Corporal in the 2nd Durham (Seaham) Artillery Volunteers, had only recently returned from the National Artillery Association competition at Shoeburyness.

2nd Durham (Seaham) Artillery Volunteers parade on the Terrace Green
Courtesy of Seaham Family History Group

One month earlier he was carried around the town shoulder high by his comrades because he had won the coveted Queen's Cup donated by HM Queen Victoria, nine silver cups and the badges of the NAA for the highest aggregate score in the 64-pdr gun competition. He was to receive his prizes from the Marquess of Londonderry at the Seaham Flower Show on the 8th September. Tragically he changed his shift and went to work on the 10.00 pm shift on 7th September so that he could attend the Flower Show the next day.

It was reported that Hindson had some premonition of his death and had told his wife some days earlier that he was working in a dangerous part of the pit. He remarked to his wife "Well Hinnie, I'm feared that if there should be an explosion, I'll be knocked about worse". He had set off twice and returned home again on that night but set off again for a third time from which he did not return. (Reynolds Newspaper, 1880). The funeral cortege left his residence in John Street, Seaham Harbour followed by his family and friends, about fifty members of the Artillery Volunteers in full dress uniform, and a large number of members of the Oddfellows Lodge. More than several hundred people witnessed his uniformed comrades in the 2nd Durham's carry his coffin into St John's Church covered in flowers and proudly displaying the Queens Cup on top. During the service Sergeant Triffit, winner of the Scotland Cup on the 64-pdr guns at Shoeburyness the year before, stood guard over the coffin with tears in his eyes. The Volunteers had lost a total of 26 NCO's and gunners mainly from Major Warham's battery in the explosion including: -

- Corporal Thomas Hindson
- Corporal Walter Murray
- Corporal Robson Dawson
- Bombardier Thomas Alexander
- Gunner Robert Potter

- Gunner Matthew Charlton
- Gunner Silas Scrafton
- Gunner John Roper
- Gunner George Roper
- Gunner Joseph Cowey
- Gunner Thomas Greenwell
- Gunner Edward Brown
- Gunner Jacob Fletcher
- Gunner Alexander Sanderson
- Gunner Anthony Greenbank
- Gunner Michael Smith
- Bandsman Thomas Grounds
- Bandsman John Grounds
- Bandsman James Johnson
- Bandsman John Miller

Later the sum of £28 10s was offered by the Marquis of Londonderry for the Queens Cup won by Corporal Thomas Hindson so that it could be put on display with the rest of the Brigade cups and awards. This was accepted by Jane the widow of Corporal Hindson who now had two young children to bring up on her own.

The families and friends of the missing men continued to congregate around the pithead waiting for any news from the explorers returning to bank. Despite the pitiful and grief-stricken scene, a most distasteful event arose. Thomas Laverick was one of the spectators looking from the gangway of No. 3 pit when he witnessed a most unseemly exhibition of callous and cold-hearted behaviour from Police Constable 268. "Move on" he said to a poor fellow who was clearly consumed with grief. "Where must I go" asked the man; "I have stood here two days and nights waiting for them to bring up my father who is killed in that pit". The policeman's response was a severe push against the man's chest and a threat that he would put him off the gangway altogether if he continued his impertinence. Angered by the scene he had witnessed Thomas Laverick wrote to the editor of the local newspaper in the hope that the superiors of the "self-important, new-fledged policeman dressed in the public money's clothing" would instruct the scoundrel that the public would not submit to that sort of treatment.

Seven days after the fatal explosion the work of clearing the debris from No. 1 pit shaft was finished and efforts began to extricate the cage. It was eventually clear of the debris and hauled by ropes and brought to bank. What remained beneath the cage was removed bit by bit until the shaft was clear. The clearance of the shaft materially improved the ventilation of the mine although it was believed that there was still a sufficient accumulation of gas in certain parts of the mine to form the elements of another explosion. The shaftsmen began to repair the damage to several fathoms of skeats and bunting within which the cage, when repaired, would slide. The colliery management eager to disperse pockets of gas in the mine consulted with several mining engineers on whether it would be safe to re-light the furnaces to secure better ventilation. It was decided not to re-light the furnaces for fear of a second explosion. The officials of the Durham Miners Association, Messrs Foreman and Patterson, requested that they be allowed to enter the mine as soon as the cage was brought back into service and to make what examination they thought fit on behalf of the men. Mr Thomas Burt the Member of Parliament for Morpeth, a lifelong labour champion and miner's leader and a member of the Royal Commission for Accidents in Mines visited Seaham Colliery to see the scene and to enquire about progress in recovering the bodies.

Thomas Burt MP for Morpeth
The first working miner to become a Member of Parliament

Mr Burt MP also took the opportunity to visit his cousin Thomas Burt the Seaham lodge official who had been asked by the DMA to join their teams inspecting the mine on behalf of the men.

The excitement and visitors to the colliery village which existed just a few days earlier had calmed down and only a few strangers could be seen wandering around the colliery rows. An eerie quietness reigned throughout the village.

CHAPTER 13

The adjourned inquest reopens

Day 8, The 15th September 1880

The Londonderry Literary Institute

The adjourned inquest on the explosion at Seaham Colliery was opened on the morning of 15th September 1880 at the Londonderry Literary Institute. There was only a very limited attendance of the public but in addition to the previous inquest members were Mr Foreman and Mr Patterson of the Durham Miners Association; their legal counsel Mr Bowey and Mr Marshall and Mr Willis, Inspector of Mines for Northumberland. Mr Corbett produced a vertical plan of the shafts, seams, faults and drifts at Seaham Colliery in order that the jury could understand the workings of the mine. (See Diagram 2 overleaf)

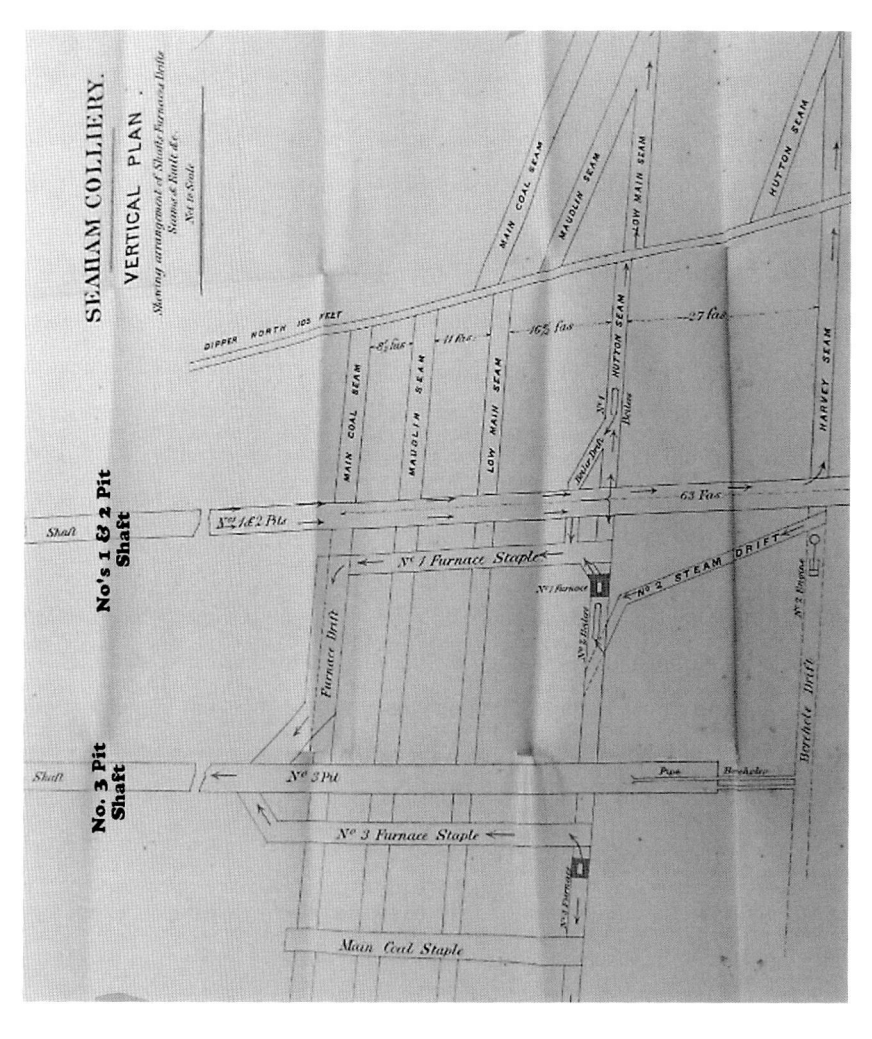

Diagram 2: Vertical Plan of Shafts, Seams, Faults and Drifts at Seaham Colliery

September 8th 1880

Courtesy of Durham County Records Office D/X 1051/6

The Coroner, Crofton Maynard, announced that he intended to examine as many survivors as thought necessary who were brought out of the pit alive so they could tell the jury what occurred in the localities they were working at the time of the explosion. The order that he would take evidence would be the hewers and other workmen first; then he would hear the overmen and finally the viewers and any scientific evidence available.

William Cummings, shifter, was the first to be called to give evidence. He was with John Mason and George Dixon in the Hutton seam, No. 3 in the south locality, known as the back of the staple. He felt a heavy wind followed by a fearful bang. He told the other two to pick up their lamps and run for their lives. There was a roof fall about thirty or forty yards from where they were and the air was very thick. Working through the fall they came across a little boy very ill. He told the Coroner "John Mason and I carried on and we expected George Dixon to follow but when we got further outbye we realised that George Dixon must have stayed with the boy." The explosion happened at 2.20 am and the ventilation seemed to be good when he started his shift. He couldn't say what was the cause of the explosion but he recalled that Thomas Hindson told him on the Monday that he had been instructed to enlarge the refuge holes using blasting powder. When Cummings progressed further, he found one man dead with some tubs against him near the main coal staple and some of the timbers were on fire. They were too exhausted to put out the fire and the staple bottom was wrecked the blast apparently having come up the staple. When he got to the shaft he saw one man badly burnt.

William Harvey, another shifter, gave evidence next. He was in the Hutton seam in a branch in the east way when he experienced a change of air which he felt to be serious so he made his way outbye. In the engine plane he noticed that the man who had been making refuge holes had gone and he progressed to the overman's cabin where he found men coming back because of the stithe up ahead.

Ralph Marley, a stoneman, was working in the main coal in the skirting way about 1,200 yards from the No. 3 shaft when he heard two shots in the air, the first was a mere whiff but the second was a severe shock and then he noticed a smell. He and his mate decided that something was wrong and a workman called Thomas Wilkinson went towards the shaft and found smoke so they travelled to the staple top where they found fresh air. The staple was used for dropping coals from the main coal to the Hutton seam. At the staple they found evidence of wreckage. They found Robert Wardle in the overman's cabin about 20 yards from the staple top. He was not burned but cut about the face. They got Wardle to the shaft and they all got out of the pit alive. No one in that section of the pit was killed except Rollins.

Mr AG Kent described the scene when he found the body of James Brown. "As my marrows and I were going between Nos. 1 and 3 shaft about four o'clock on the morning of the explosion, we came upon the body of James Brown lying in a refuge hole surrounded by fire composed of burning wood. I drew the corpse out and with my own hands and feet dashed a quantity of the terrible pit dust upon the fire. It completely extinguished the fire."

A number of other men who survived the explosion and were working in the main coal or the Harvey seam were called to give evidence. Robert Wardle the night shift Deputy; George Thompson, stoneman and Thomas Johnson, shifter all told the jury where they were when the explosion happened and what they saw and did from that point until they were rescued and brought out of the pit.

The inquest was then adjourned until Tuesday the 19th October, 1880.

CHAPTER 14

Exploration continues into No. 1 Pit

Day 9 to 10, 16th to 17th September 1880

The Manager Thomas Stratton had expected to put the cage in on No. 1 Pit shaft today but the work of putting in new skeats and bunting had not been completed. In charge of the work in the No. 1 pit shaft was Mr Coulson the well-known County Durham sinker and son of Mr William Coulson who had charge of repairing the shaft after the Hartley Colliery disaster in 1862. Meanwhile, men had been descending No. 3 pit shaft and passing underground to No. 1 pit to continue exploration. No more bodies had been found but the officials expected that the area they were exploring would soon reveal a considerable number of bodies. In anticipation of this a second large trench had been dug in Christchurch graveyard close beside the trench in which fourteen of the victims of the explosion were buried the previous Sunday.

The Bishop of Durham was becoming concerned for the plight of the families, the widows and children left destitute by the disaster. In a letter to his Archdeacons, he recommended that collections for relief be made in every church and suggested that sums collected be entrusted to the Seaham Relief Committee of which the Rev. WA Scott was Chairman. After relieving the needs of those sufferers for whom sufficient provision was not made in other ways he stated that the committee should eventually hand over any residual monies to the Northumberland and Durham Miners Permanent Relief Fund. The Permanent Relief Fund was doing great and benevolent work for miners in Northumberland and Durham but the catastrophe at Seaham would prove to be a crippling blow and drain on its resources.

By Friday morning the weather had improved from the gloom of the previous two days and a large crowd at the pithead anxiously stood in bright, clear sunshine. The news had reached bank that bodies had been found. The exploring party had reached the "far off way" in the Hutton No. 1 seam and they had discovered seventeen bodies. None of the bodies had been mutilated or burnt and therefore must have succumbed to the effects of afterdamp. They were found in close proximity to each other in various positions and the explorers described the sad scene. One man was in an attitude of rest leaning against a wall; another was lying on his face; another sitting and another had his mouth open as if gasping for breath. No efforts were made to identify the bodies before they were sent to bank. A note was made where each body was found. A number was given to each body which was marked in chalk on the coffin and sent to bank. The corpses were then taken to the identification room of the dead house where the lid was taken off and the body sufficiently exposed to permit the relatives of the men to identify them. The distorted but undamaged bodies of the following men were identified: -

Thomas Greenwell, Deputy
Robert Graham, Hewer
Charles Horan, Hewer
George Shields, Hewer
John Short, Shifter
Matthew Charlton, Hewer
David Williams, Putter
Thomas Forster, Shifter
William Richardson, Hewer
John Jackson, Shifter
Frank Watson, Shifter
Roger Henderson, Shifter

Michael Henderson, Putter
Dominic Gibbons, Shifter
William Henderson, Shifter
James Slavin, hewer
Michael Owens, driver

Mr Hall of Haswell who was in charge of one of the exploration parties pushed further into the Hutton seam throughout the afternoon. The exploring parties were joined by Mr Foreman, Patterson, Crozier, Banks and Burt from the Durham Miners Association who were representing the men. On returning to the surface Mr Foreman reported that he went into the pit at 12.00 midnight and came out at 6.00 am in the morning. He visited the Maudlin stables; the bottom of the staple in No. 1 Hutton seam; the No. 3 furnace of the upcast shaft and No. 1 Hutton seam where they joined Mr Stratton and his exploration party. Ventilation had been restored as far as the Polka Way. When he left to come back to bank the second exploring party was trying to get into the "third east way" where bodies were expected to be found. He was highly satisfied with all the work he witnessed in his shift. He found the smell of dead horses was very strong but the men had begun disinfecting down the areas affected. The Maudlin seam was not expected to be explored until the No. 1 Hutton was cleared. He commented "The energy with which Mr Stratton works underground is astonishing. He went down with the same shift as me and when I came away he was still in the mine."

Thomas Burt the local lodge official who joined the exploration parties as one of the DMA representatives also made a report. His shift started at midday and his party travelled about a mile away from the shaft to the third east way where they found the first body. He was found in the flat about 200 yards from where he had been working. Thomas Burt commented "He was terribly swollen with the effects of afterdamp. Beyond was his marrow and two lads but his party went no further from these bodies. We went to the southern district to the far-off way where the tubs were considerably smashed by the effects of the explosion and then progressed to the landing and found a man who had made his way past the broken tubs but had fallen with his lamp in one hand and the other under his brow. He had died from the afterdamp. Just a few yards further another man was found almost naked and then a short distance further a group of nine men were found in a group, some lying and some sitting. They seemed as though they were asleep."

CHAPTER 15

Messages from the dead

Day 11 to 14, 18th to 21st September 1880

The work carried out by the men down the mine had been twofold in the last couple of days. Efforts focused on recovering all of the bodies accessible from the No. 1 pit and secondly to improve the ventilation in the mine. The body of opinion amongst the mining engineers was that it would be too dangerous to light the furnaces and an alternative scheme was adopted to draw air through the mine workings. A large tree branch was hung over the mouth of No. 2 pit downcast shaft and a steady stream of water was released over it. A steam pipe was taken down 240 feet into the No. 3 pit upcast shaft and allowed to escape. The theory behind this scheme was that the water descended in drops like rain and cooled the air in the downcast shaft whilst the air in the upcast shaft was heated by the release of steam which rendered the heated air lighter. This drew the air through the mine with greater velocity.

The exploration teams had spent the previous two days trying to clear away extensive roof falls in No. 1 Hutton seam. Eventually the falls were broken through and after proceeding a very short distance Mr Stratton and Mr Bailes junior discovered some bodies lying in groups of two, three and four. It seemed that they had run out after the explosion until they came across the roof fall and then were struck down with the afterdamp and perished. Near the bodies of a group of four they found a piece of brattice board about 2½ feet long and 2 inches wide. Written on one side in chalk in the same handwriting were the names of the men. They were: -

John Riley, shifter

James W McLoughlin, chock drawer

Jacob Fletcher, shifter

Richard Driver, packer

On the other side were the words: -

"Five o'clock: we have been praying to God"

Thomas Stratton instructed the exploration team to take the bodies immediately to the shaft so they could be taken to bank. The whole of the No. 1 Hutton had now been explored and all the bodies in that part of the mine had been recovered. All efforts were now concentrated on clearing away the fall in the 3rd East Way. Almost 400 tons of stone were removed from that fall. The work progressed faster than Mr Stratton had anticipated and about 150 yards beyond the fall the explorers came across more bodies of men who had died of afterdamp. They were: -

- George Lamb, stoneman
- Henry Ramsey (aka Ramshaw), shifter

Ramsey and Lamb had been working together and their bodies were found in a refuge hole. At the time of the explosion, they were having their bait part of which was projecting from Ramsey's jacket pocket. The exploring party were taken by surprise when they came across the body of Henry Ramsey. He was better known as Ramshaw and he had served in the Afghan campaign. After an injury to his leg he was invalided out of the army and he had only worked three shifts in the pit when the explosion happened. However, his body had previously been identified by his aged father from amongst the first batch of bodies some days earlier and that body had been buried by the family. This body was unmistakeably Ramshaw as he was wearing army boots and a distinctive straw belt which he was wearing when he left to go to work.

A short distance beyond Ramsey and Lamb were another group of bodies who had died in a variety of positions, some on their backs, some on their sides and a number resting on their knees with their faces close to the earth as if they were desperately trying to breath in any air lingering below the afterdamp. All of the men had their lamps clenched firmly in their hands or lying beneath them. Some of them were recognisable but they were swollen and in an advanced state of decomposition. The bodies belonged to: -

- Thomas Hayes senior, stoneman
- Thomas Hayes junior, stoneman
- Robert Haswell, putter
- John Lonsdale, hewer
- William Hood, shifter
- Mark Philips, stoneman
- William Potts, deputy
- Luke Smith, hewer
- James Clark senior, hewer
- James Clark junior, hewer
- Alex Sanderson, stoneman
- William Fife, stoneman
- William Barrass, hewer
- James Hedley, putter
- George Hopper, shifter
- Joseph Birkbeck, shifter
- Robert Straughan, putter
- Michael Keenan, shifter
- John Knox, driver

- George Page, stoneman
- John Lock, stoneman
- John Kirk, shifter
- David Knox, driver
- Anthony Scarfe, brakesman
- Robert Clark, shifter
- George Norris, shifter
- Charles Dawson, master shifter
- William Bell, shifter
- Thomas Smith, shifter
- Edward Burns, putter (AKA Pinkard)
- George Roper, hewer
- Michael Henderson, shifter

The news had spread that more bodies had been found and were being brought out of the pit. Crowds of men and women began to linger around the dead house as the bodies were taken in. Groups of three and four from each family were allowed into the dead house to claim the bodies of their missing loved ones. A notice from the Rev. WA Scott posted at the dead house reminded the mourning families "I earnestly recommend that the bodies of our bereaved friends should not be kept too long after their recovery and identification and that funerals on Sundays be avoided as far a possible." One poor old man called Waller had walked incessantly day-after-day from Seaham Harbour to the pit in the hope each time of recovering the body of his son. The sixteen-year-old putter Joseph Waller was not found until almost one year later.

Robert Hartley, deputy and one of the working parties with Mr Stratton commented on the scene when they found the four miners that had written on the brattice board. He said "I followed Mr Bailes to where the poor fellows were lying. The sight of them was most affecting and I do not think there was a man among us that wasn't moved by the sorrowful picture before us. The first man we came to was Jacob Fletcher, stoneman, who was lying with his head over on his right side and his knees drawn up. John Riley was lying across the way at the feet of Fletcher drawn up in much the same position. Richard Driver was sitting with his head and shoulders against the side of the coal. James McLoughlin was sitting beside Driver. There were three lamps standing together in a row between Fletcher and Riley. Fletcher's lamp was standing close to his breast. The bodies were not at all disfigured only very swollen. Their faces were black but their countenance had not changed and they seemed as if they had become exhausted and succumbed to the afterdamp. The lamps must have been burning at the time the one man wrote on the piece of board as it was written in a good hand. The lamps were supplied with oil for twelve hours and would have burned until all of the oil was consumed"

Twenty-five of the ponies saved in the main seam were brought to bank up No. 3 pit shaft. New guides were being fitted in the shaft and preparations were made to insert a new cage. All of the exploration teams had been working through No. 1 pit shaft as a continuous flow of water descended down No. 2 pit shaft to assist the ventilation.

The total number of bodies recovered was now 88 and there was believed to be another 76 undiscovered in the mine all in the Maudlin seam. Exploration of the Hutton seam was discontinued after all of the bodies were taken to bank and all efforts were concentrated on breaking into the Maudlin seam.

CHAPTER 16

The Relief Fund

Five days after the dreadful explosion on the 8th September 1880 a meeting was held in the Londonderry Literary Institute to discuss the plight of the poor widows and their children who were now destitute because of the loss of their menfolk. It was resolved that a "Subscription List" be opened to assist in the relief of the bereaved families in conjunction with the Northumberland and Durham Miners' Permanent Relief Fund.

A proposal by Viscount Castlereagh was agreed that a local committee be appointed to collect subscriptions which would be eventually handed over to the Permanent Relief Fund. However, the committee would have the power to reserve such sums out of subscriptions received and to pay out such amounts it thought fit to sufferers that did not come under the provisions of the N&DMPRF. An influential committee was formed including: -

- Rev. WA Scott, Chairman
- George Young, Secretary
- Mr Harper of Woods & Co Bank, Treasurer

Committee members included The Marquess of Londonderry, Viscount Castlereagh, JB Eminson, VW Corbett, Messrs Ditchfield, Stratton and Warham, Local clergy of various denominations, Messrs Howie and Blythe of the N&DMPRF and most of the influential inhabitants of the town.

Her Majesty Queen Victoria sent a cheque for £100 towards the relief fund when she heard of the disaster. The appeal for contributions to the Relief Fund was circulated throughout Great Britain and a list of subscribers and the amounts donated was circulated in the local newspapers as shown below: -

The following Contributions have already been received or promised:—			
The Marquess of Londonderry	200	0	0
The Marchioness of Londonderry	50	0	0
Lord Viscount Castlereagh	50	0	0
Viscountess Castlereagh	25	0	0
Messrs J. and H. H. Thompson	10	10	0
Messrs Fletcher, Son, and Fearnall	10	10	0
The Sunderland Brigade Depot, per Lord John Taylour	7	5	9
Laura, Countess of Antrim	5	0	0
F Le Sucur and Son	5	0	0
Robert Bardfield, Esq., Sheffield	5	0	0
A. T. Crow, Esq., per Rev. A. Scott	2	0	3
Sir George Elliot, Bart.	105	0	0
R. L. Pemberton, Esq., & Mrs Pemberton	25	0	0
Haggie, Bros.	25	0	0
Wm. Jackson, Esq., p Mayor of Sunderland	21	0	0
Messrs Woods and Co.	105	0	0
J. L. Wharton, Esq.	25	0	0
Anonymous	100	0	0
Henry Ritson, Esq.,	3	3	0
Robert Oldrey, Esq., Wales	10	10	0
J. R. Eminson, Esq.,	10	0	0
R. Brydon, Esq.,	5	0	0
W. Pallister, Esq.,	5	0	0
E. A. Philips, Esq.,	1	1	0
R. Thorman, Esq.,	10	0	0
Thomas Chilton, Esq	5	0	0
Collection Boxes	13	17	5
Thomas Scawin, Esq	10	10	0
Rev. A. B. Grimaldi, formerly Curate of New Seaham	1	1	0
G. R. R., Bath	1	0	0
Calder Bros.	10	10	0
McIntyre Bros., and Co.	10	10	0
R. Thompson, London	5	0	0
Matteson and Chapman	10	10	0
Joseph Cook and son	10	0	0

	£	s	d
H. Tonkinson, Esq.	2	2	0
Duchess of Marlborough	25	0	0
R. J. T. Todd, Esq., London	20	0	0
Messrs. D H. and G. Hagrie	5	0	0
Captain John Coates, Newcastle	1	0	0
Messrs J. P. Austin and Son	10	10	0
Lady Adelaide E. C. Law	21	0	0
Martin Wiener, Esq.	25	0	0
W. J. Young, Esq., J.P.	5	0	0
Rev. A. Bennett	2	2	0
Messrs N. Willis and Son	2	2	0
N. and J. Taylor, Sunderland	10	0	0
Wm. Lough, Esq	3	3	0
Mr T. W. Johnson	0	2	6
A Friend—"G.P."	0	10	6
Dr F. Sherwood Stoeke	5	0	0
Messrs Joseph Crawhall and Sons	10	10	0
H. B. Wright, Esq.	10	0	0
A. G. Boulton, Esq.	3	3	0
A Friend, per do.	0	10	6
Rev Jas. Colling	10	0	0
Fitters at Messrs John Abbot & Co.'s Gateshead	1	4	1
Rev. J. D. Middleton, West Cowes	1	1	0
"B." Newcastle	2	2	0
Dr. C. Gibb, Newcastle	5	5	0
"T. W.," Newcastle	2	2	0
B. F. Bolman, Esq., Newcastle	2	0	0
Mr Alex A. McLaurin	0	5	0
Part proceeds of Bazaar at Castle Eden, per Rev. J. Burdon	50	0	0
"T.C B.," London	2	2	0
Messrs Hill, Wood, and Co., London	52	10	0
,, Harris and Dixon, London	52	10	0
,, W. Cory and Son, London	25	0	0
Charles Milnes, Esq. London	25	0	0
Messrs Green, Holland, and Sons, London	10	10	0
,, Rickett, Smith, and Co., London	10	10	0
,, Hinton and Horne, London	10	10	0
,, Beadle Bros., London	10	10	0
,, F. Longstaff and Sons, London	5	5	0
,, Charrington, Sells, Dale, and Co., London	5	5	0
,, G. J. Cockerell and Co., London	5	5	0

	£	s	d
Thomas Parker, Esq	2	2	0
James Hartley, Esq	100	0	0
Tannaker Buhicrosan, Esq	2	2	0
R. Smey, Esq.	10	0	0
H. R. Lempriere, Esq., London	10	10	0
James Brooks, Esq., London	3	3	0
Dowager Lady Barrington	25	0	0
Lambert Brothers, London	52	10	0
A. J. Cavel, Esq.	25	0	0
Captain Going	0	1	0
Mr John Nicholson	5	5	0
P. C. Mann Esq., J.P.	20	0	0
Walkers, Parker, Walker, and Co., Newcastle	10	0	0
R. J. Walker, Esq., Newcastle	10	0	0
Ernest H. Walker, Esq., Newcastle	2	2	0
Captain Lancelot A. Gregson, Wimbledon	10	0	0
Members of Yearly Fund, held at Mr Heckles', Foresters' Arms	1	0	0

ADDITIONAL SUBSCRIPTIONS.

	£	s	d
Thos. Douglas, Esq.	10	0	0
Nichlas Stevenson, Esq.	1	1	0
Offertory at Auckland Castle	4	3	6
Messrs Alexander and Wood	1	1	0
Stanners Close Steel Works, Wolsingham	5	5	0
Mrs Mary Thompson	1	1	0
Miss Carrie Thompson	0	10	0
Mr Jno. Wilkinson	0	5	0
Major McKenzie	3	3	0
"T.R.W."	5	5	0
"J.L.M."	5	0	0
Anonymous	5	0	0
Rev. Mr Harrison	5	0	0
T. Bell, Esq.	5	0	0
Seaham Branch Colliery Enginemen	5	0	0
Geo. Blagdon, Esq.	5	5	0
W. Dowber and Son	10	10	0
J. R. Upton, Esq	10	10	0
Proprietors of *Durham County Advertiser*	5	5	0
Col. Scurfield, Hurworth-on-Tees	25	0	0
O. A. Bushell, Esq.	1	1	0
Capt Horlock, s s. 'Viscount Castlereagh'	5	0	0

					£	s	d
,,	Gammon, Son, and Carter, London		..		5	5	0
,,	T. S. and Co., Parry, London		..	..	5	5	0
,,	Lea and Co., London..	..	..	..	5	5	0
,,	Pope and Gill, London		..	—	5	5	0
,,	Brewis Bros., London	—	..	—	5	5	0
,,	Scott and Dinham, London	..	..	—	5	5	0
J. Hazell, Esq., London			..	—	3	3	0
Messrs W. Dowell and Co., London..			..	—	3	3	0
,,	Corrall and C., London		..	..	2	2	0
,,	J. G. Bryan and Co., London		..	..	2	2	0
,,	Bay and Sons, London		..	—	2	2	0
,,	Charles Hart and Co., London		—	—	2	2	0
,,	Wm. Lee, Son, and Co., Halling, Kent			..	2	5	0
Captain Huntrod		..	..	..	1	0	0
,, Lonsdale		..	—	..	1	0	0
,, Slater		..	..	..	1	0	0
Mr N. Meek		..	..	..	0	5	0
Mr E. Wilkinson		..	..	..	0	3	0
Firemen of Harvest Queen's			..	..	0	2	0

Further Contributions will be thankfully received by the Treasurer above-named, and at any of the Branches of Messrs Woods and Company's Banks, and also at the Union Bank of London.

Murton Colliery held a meeting and resolved that each man should contribute one shilling and each boy sixpence to the Relief Fund. South Hetton Mechanics contributed sixpence per man. Over the next few weeks many more subscribers came forward until the total funds received by the committee was in excess of £13,000. Of this sum £7,500 was invested with the River Wear Commissioners to deal with current and urgent cases of need as determined by the Relief Committee and over £4,000 was handed over to the N&D MPRF to top up there funds which had been severely drained by the effects of the disaster.

Permanent Relief Funds were distinct from the voluntary subscriptions and donations received from the public for disasters such as the Seaham Colliery Disaster. Permanent Relief Funds were established in many coalfields in the second half of the nineteenth century. They were intended to provide benefits to the widows and 'orphans' of miners killed in their work. They were 'permanent funds' because, unlike friendly societies or other agencies, they paid benefits indefinitely. The Northumberland and Durham Miners' Permanent Relief Fund was founded in the wake of the Hartley Colliery disaster in 1862 which was the first such permanent relief fund. More than £85,000 had been raised nationally from public appeals towards the relief of the families of the 204 men killed in the disaster. This sum was much more than was required to provide long term relief to the widows and children. However, the inequality of relief was expressed by many miners in the northern coalfields who knew of many instances of individual deaths that did not attract such large subscriptions as in the case of major disasters. It led to calls for permanent long-term support instead of ad hoc fundraising. This was the catalyst for the formation of the Northumberland and Durham Miners' Permanent Relief Fund. In 1863 surplus funds from the Hartley Relief Fund were transferred into the N&D MPRF. In addition, weekly deductions of 1p were made from the weekly pay of miners who were full members of the scheme and 1/2p from boys who were half-members. Coal owners made a voluntary contribution of a percentage of the miners' subscriptions into the fund. Prior to the explosion of 1880 Lord Londonderry had paid into the N&DMPRF about 20% of the contributions paid in by Seaham Colliery members.

A notice was posted outside the temporary offices of the Miners Permanent Relief Fund four days after the explosion on 12th September 1880. The officers of the fund were aware that some families would shortly be in need of funds to bury their menfolk. The notice stated that burial money would be paid immediately to the relatives of those whose bodies had been recovered. The legacies and fortnightly allowance paid out from the Permanent Relief Fund to widows and children were based upon the following scale: -

Accidental death of unmarried member under 18 years ….. £12
Accidental death of member over 18 leaving no dependant relatives …. £23
Accidental death of members over 18 leaving dependant relatives ….. £5

Widows or other dependant relatives …………… 5 shillings per week

Each child of members accidentally killed ……….2 shillings per week

Tragically the widow of Anthony Ramshaw who was killed in the explosion died broken-hearted shortly after the burial of her husband. The N&D MPRF did not make allowances for the burial of members' widows and the Seaham Colliery Relief Committee granted the family the sum of £5 from the discretionary funds available to them.

The calculation of benefits to widows and children paid out from the Northumberland & Durham Miners Permanent Relief Fund are listed in Appendix 1 attached.

Some miners at Seaham Colliery were also members of friendly societies and received benefits from those schemes. Walter Murray, one of the gunners who had attended the National Artillery Association big gun competition at Shoeburyness with Corporal Hindson had his life insured with the Gresham Life Society for £100. Although his body had not yet been recovered the directors dispensed with the usual formalities and forwarded a cheque to his widow for the full £100. Fifty-nine of the victims were members of the Prudential Assurance Company with headquarters in Foyle Street, Sunderland. The Prudential proudly announced that they had promptly paid out the benefits from the policy amounting to more than £600 within three days of the date of the disaster even to families where the body of the named policyholder had not yet been recovered. Unfortunately, five policies had lapsed because the miners had been unable to pay the premiums in the spring of 1880 and those policies were null and void. Despite this the Prudential paid a gratuity to the dependants in recognition of the special circumstances. The miners who lost their lives and were members of the Prudential Assurance Company are shown on the statement below.

PRUDENTIAL ASSURANCE COMPANY.

List of Claims paid by the above Company in connection with the

Seaham Colliery Explosion.

Superintended by Mr. R. D. HILL, 22, Foyle Street, Sunderland.

The prompt and handsome settlement of all claims by the Prudential Assurance Company, to the amount of upwards of £600, within three days after the sad disaster at the above pit.

Name.	Address.	Amount.	Name.	Address.	Amount.
Butler, John	Bottle Works, Seaham Harbour	14 14 0	Lonsdale, John	Seaham-street,	9 1 0
Owens, Michael	Back Sa. Railey-st., "	9 16 0	Jackson, John	"	9 9 0
Johnson, Edward	Frances-street, "	6 6 0	Hutchinson, Thomas	"	6 6 0
Cank, Joseph	Henry-street, "	11 4 0	Redding, John W.	Cornish-street,	10 2 0
" Joseph	"	15 4 0	Taylor, William H.	California-street,	5 11 0
Bowran, Edward R.	Hall-street, "	9 11 0	Lonsdale, Joseph	Hall-street,	19 16 0
Hays, Thomas	"	10 3 0	Potter, Robert	Model-street,	14 11 0
" Thomas	Church-street	7 7 0	" Robert	"	9 11 0
Veitch, William M.	Cornish-street	9 16 0	Serafton, Silas	Vane-terrace,	5 12 0
" Samuel	"	9 11 0	Greenwell, Thomas	School-street,	11 4 0
Henderson, Michael and William	Cook-street	13 4 0	Shields, Robert	"	20 4 0
" Roger	"	11 2 0	Brown, James	Seaham-street,	12 12 0
" Michael	"	10 1 0	Michael, Keenan	Henry-street,	10 0 0
Dixon, Leo R.	Australia-street,	7 7 0	Hood, William	"	15 16 0
Gibbon, Dominic	Vane-terrace	10 2 0	Lawson, Robson	Calliornia-street,	9 16 0
Dunn, Robert	Seaham-street,	7 18 0	Cole, Richard	Summerson's Buildings, Seaham Harbour	12 12 0
Neasham, John	California-street,	8 10 0	Turner, Alfred Jas.	Cornish-street,	22 1 0
Williams, Thomas H.	Cook-street,	10 14 0	Dawson, Walter	Henry-street,	8 14 0
Wilkinson, William	Cornish-street	17 12 0	Berry, William	California-street,	11 2 0
Keenan, Thomas	"	6 7 0	Morton, William	William-street,	5 11 0
Forster, Thomas	"	10 10 0	Crossman, William	Australia-street,	8 11 0
Clark, Thomas	Doctor-street,	8 16 0	Short, John	Church-st., Seaham Harbour	Gratuity 2 7 0
" James W.	"	11 3 0	Redshaw, Benjamin	White's Yard,	Gratuity 5 15 0
" James W.	"	8 10 0	Smith, Luke W.	California-st., Seaham-Colly	Gratuity 5 11 0
Slaughter, Robert	"	3 1 0	Grounds, Thomas	Butcher-street,	Gratuity
Batey, John	Butcher-street,	8 4 0	Simpson, William	California-st., "	
Morris, William	William-street,	6 2 0	Grounds, John	"	
Best, James	William-street, Seaham Colliery	4 8 0			£620 15 3

In the five latter cases the Policies were lapsed, during the spring of the year, but although the Policyholders allowed their Policies to become null and void, in consequence of not being able to pay the premiums, the Company paid the Gratuities as stated, and in many instances where members had just newly entered, the claim was paid in full.

SEPTEMBER, 1880.

Statement from Prudential Assurance Company of policies paid out to families of the victims of the Seaham Colliery Explosion Courtesy of Durham Records Office (DRO D/Fn/209/1)

At a full meeting of the workmen at Seaham Colliery on 18th September the issue of the plight of miners out of work with no wages because of the explosion was discussed. It was agreed to establish a relief fund to relieve the distress caused by the explosion not only for the bereaved widows but also the distressed families of men out of work who had no claim on the Permanent Relief Fund. This relief fund was announced in the local press and many subscriptions were received over the next few weeks.

Quite by coincidence later that same year the government passed the Employers Liability Act of 1880. The Act required employers to be statutorily responsible to compensate employees for accidents in the workplace. In a blatant attempt to abdicate their responsibility for workplace accidents the coal owners proposed to "contract out" of the Act subject to paying additional contributions into the N&D MPRF. In other words, they attempted to avoid the employer's liability to compensate for accidents by making the N&D MPRF assume their compensation liabilities. The miners refused to have anything to do with this and the coal owners were bound by provisions of the Employers Liability Act.

The impact of the Seaham Colliery explosion on the N&D MPRF was significant. The 164 victims left 107 widows, 2 mothers, 2 guardians and 249 children dependent on the Fund. Four years after the Seaham Colliery explosion the reports of the N&DMPRF showed that the costs borne by the fund of the 1880 catastrophe had been £9,000 and the annual liability on the fund was £1,700 per annum. Payments for children continued until either a boy reached the age of twelve or a girl reached the age of fourteen so the last payment from the fund for children would have been no later than 1894. Payments to widows continued throughout widowhood and ceased upon re-marriage. By 1920 there were still nine widows supported from the N&D MPRF as follows: -

Relief Started	Name of Widow		Born	Relief Ended	Reason Ended
8/9/1880	Alexander	Isabella	1844	13/01/1929	Death
8/9/1880	Dawson	Jane Anne	1848	02/01/1928	Death
8/9/1880	Dawson	Ellen	1851	26/03/1925	Death
8/9/1880	Murray	Elizabeth	1839	15/12/1920	Death
8/9/1880	Potts	Ellen	1835	09/10/1921	Death
8/9/1880	Rawlings	Margaret Ann	1841	13/04/1921	Death
8/9/1880	Shields	Mary Ann	1857	02/10/1937	Death
8/9/1880	Smith	Margaret	1847	12/09/1929	Death
8/9/1880	Venner	Joanna	1835	24/11/1920	Death

The very last payment to widows of the men who perished in the 1880 explosion was made to Mary Ann Shields who died on 2nd October 1937 fifty-seven years after the death of her husband James Shields.

CHAPTER 17

Explorers find more messages from the dead
Day 15 to 22, 22nd to 29th September 1880)

The exploration teams came across a heavy fall in the Maudlin seam caused by the explosion. The extent of it could not be defined with any certainty but it was believed to be not less than forty yards in length. The fall covered so much of the workings that the men could not see over it and it was expected that there would be many more obstructions to contend with as they began clearing away the debris and stone.

A meeting took place in the colliery office with mining engineers Messrs Hall, Lishman and Forster to consider progress over the previous two days and also to confirm arrangements for lighting the furnace in No. 1 Hutton. Although the furnace was lit it soon had to be extinguished because blackdamp was being forced through the seam where the exploration parties were working and extinguishing their lamps. However, it was hoped that the furnace could be re-lit within the next few days.

Over 500 tons of debris was removed from the great fall in the Maudlin seam and it was believed that a large quantity still remained to be cleared and more roof falls were expected to be encountered. The men worked vigorously over a number of days but more roof falls caused delays in progress. The debris from these falls was brought up No. 1 pit shaft and taken away in wagons. Meanwhile the work to repair No. 3 pit shaft continued and was nearing completion.

Eventually the working parties progressed to the foot of the Maudlin incline where they found the bodies of Edward Hall and Joseph Lonsdale. One of them was on his knees with his head bent down to the ground. On the back of his coat collar was pinned a piece of paper written by the master shifter Walter Murray stating that the men died at 3.30 am and that other men inbye were still alive at 9.30 pm up the incline. Further along the incline they found the body of a third man, William Morris and a little further on the body of the master shifter Walter Murray lying on his back with his stick in his hands and lamp by his side. The exploring party were certain they would find the bodies of more men at the top of the incline but they detected gas and returned to bank. A further working party was sent down to make a road through the devastation that they had pushed through so that the latest bodies found could be recovered and brought to bank. The number of bodies now recovered was 96 with another 68 remaining in the mine.

The repairs to No. 3 pit shaft were completed and the cage was finally put in and the dead ponies found earlier were brought up in the cage wrapped in canvas sheets. They were in an advanced state of decomposition and the stench was terrible. Over 100 ponies were carted away to a large trench that had been dug about a quarter of a mile away in a nearby field. There are some people in society that will always try to make capital out of tragedy. Mr Scott, jeweller of High Street West, Sunderland began selling some macabre mementos of the Seaham Colliery disaster. He was selling inkstands made out of the hooves of ponies killed in the disaster each one bearing the inscription "This hoof is from one of the 181 pit ponies killed in the Seaham Colliery explosion where 164 men and boys lost their lives – 8th September 1880."

The exploration into the Maudlin continued up the incline. Very soon the working party came across the body of Robson Dawson a pillarman. Near to Dawson they made the gruesome discovery of two human feet but there was no other body nearby. They were brought to bank in a wooden box. Two more bodies were found on the incline. They were John Batey, deputy overman and Patrick Carroll, hewer and neither of the bodies were mutilated. It was presumed they suffocated with the blackdamp. Pushing further up the incline the working parties came across more than twenty bodies. Most of them were found close together and they had appeared to have done all they could to prolong their lives. It appeared that the gas had been seeping from the returns into the seam where they were trapped. They had placed canvas over the apertures with stones placed over them to hold them in place. This had probably extended their lives by several hours more before they were suffocated by the afterdamp. These men were: -

- Joseph Clark, stoneman
- William Berry, packer
- Christopher Smith, shifter
- Thomas Keenan, packer
- Edward Brown, hewer
- George Diston, shifter
- Joseph Cook, hewer
- Robert Johnson, packer
- George Brown, shifter
- Robert Wharton, shifter
- John Potter, shifter
- Thomas Ground, packer
- John Thomas Miller, hewer
- James Best, stoneman
- John Ground, packer
- John Dinning, packer
- James Shields, hewer
- Isaac Ditchburn, packer

- John Owen, driver
- Joseph Lonsdale, shifter
- William Wilkinson, shifter
- James Higginbottom, shifter
- Richard Defty, packer
- William Taylor, packer
- William Sawey, packer
- John George Roper, hewer
- Thomas Cassidy, hewer
- John Sutherland, chock drawer
- Robert Shields, shifter

To the frustration of the exploring party further ingress up the three-quarter mile incline was stopped because the presence of gas was detected.

The bodies of the recently discovered victims were very quickly interred in Christchurch and St John's Church graveyard. In addition to the grief and heartache already felt in the houses at Seaham Colliery it was announced by the health authorities that typhoid fever had broken out in the village. The transfer of the badly decomposed bodies of victims to their homes to lie in open coffins for family and friends to pay their last respects had caused a typhoid outbreak. As a consequence, the Sanitary Authorities took measures for the instant internment of any bodies brought out of the pit.

Quite by surprise another exploration party discovered eight bodies in the straight down way of the Maudlin seam. They were: -

- John Watson, shifter
- Richard Cole, shifter
- Michael Smith, shifter
- William Hancock, putter

- Joseph Bowden, Hewer
- Robert Dunn, shifter
- William Moore, shifter
- Thomas Roberts, pillarman

A deal board six feet long and six inches wide was found with the bodies bearing the following statement *"The Lord has been with us, we are all ready for heaven. Ric Cole. Half-past two o'clock Thursday"*. A second message not quite so clear read *"Bless the Lord we have had a jolly prayer meeting, every man ready for glory, Praise the Lord. Signed Ric Cole"*

This board was on display in the colliery office at Seaham Colliery for many years and is now proudly exhibited in Christchurch. The display also includes a photograph of Richard Cole who wrote the poignant message.

Deal board with chalk message written by Richard Cole displayed in Christchurch
Courtesy of Alan Charlton

It is thought that the day mentioned was a mistake as the engineers thought it would be impossible for the lamps to be still burning on the Thursday. The writing is clear and it must have been written in good light and therefore presumed to be written on the Wednesday. The lamps were supplied with sufficient oil to last 12 hours but if the men had "pricked down" their lights to conserve their oil the lamps could have burned for much longer than 12 hours.

Photograph of Richard Cole displayed in Christchurch
Courtesy of Alan Charlton

A door was also written on and brought out of the pit with the words written in chalk: -

"All alive at three o'clock. Lord have mercy upon us; Together praying for help at three o'clock – Robert Johnson"

A further written record was found. A message had been scratched with a nail on a tin water bottle belonging to Michael Smith. It was found close to his body. On one side of the bottle the pitiful message to his wife read: -

"Dear Margaret – There was 40 of us altogether at 7.00 am. Some were singing hymns, but my thoughts were upon my little Mick. I thought that him and I would meet in heaven at the same time. Oh, dear wife, God save you and the children, and pray for myself. Look at the bottom."

On the bottom of the bottle were scratched the following words: -

"Dear Wife – Farewell. My last thought 'bout you and the children. Be sure and learn the children to pray for me."

Then on the other side of the bottle was scratched: -

"Oh, what a terrible position we are in – Michael Smith, 54 Henry Street"

Water bottle with message from Michael Smith - Courtesy of Alan Charlton

Michael Smith was an educated and talented man. He was the son of a schoolmaster and he played the violin. He encouraged his children to study and love music. A night of entertainment organised in the Catholic Institute of St Mary Magdalene's Church in May 1890 featured a violin solo played by his talented fourteen-year-old son. Sadly, Michael knew that his other fourteen-month-old infant son Michael was gravely ill and close to death when he left home to go to work on 7th September. His son Michael died the next day - the same day as the disaster happened. When it became known that a bottle had been found a reporter from the Illustrated London News knocked on the door of Michael's widow in Henry Street and asked if they could take away the bottle so that a drawing could be made for the newspaper. The message meant so much to Mrs Smith she refused to let the reporter take away the bottle. Instead, an artist from the newspaper sat in the front room of Mrs Smith's house and drew a sketch of the bottle. Later an offer of £50 to purchase the relic was refused by Mrs Smith.

Photograph of Michael Smith – Courtesy of Alan Charlton

The total number of bodies recovered was now 135 and the exploration continued in search of the final 29 victims of the explosion. Despite the accumulation of gas in the incline the explorers were determined to find the bodies of the other men mentioned by Michael Smith on his water bottle. He had scratched "There was 40 of altogether at 7.00 am". In order to get up the incline and to dissipate the gas every effort was made to pump as much "good air" into that part of the mine. After almost eighteen hours continuous effort during which time Mr Stratton the manager and Mr Patterson and Greenwell, miner's representatives worked continuously without rest until they reached the top of the Maudlin incline and made a thorough examination along the whole distance. There was not another body to be seen. The master shifter must have directed those missing men to another part of the Maudlin seam before the explosion occurred. The only thing they found was a piece of brattice deal nailed to one of the props with the initials of three men upon it and the time. It read: -

"W.M, I.S, R.B. All alive at 8.00 o'clock, 8th"

This piece of brattice was written almost eighteen hours after the explosion. By now the level of gas was becoming too dangerous and the exploration parties had no option but to retreat. It was agreed that no further exploration would be made in that district until the furnaces were re-lit and the ventilation was much improved.

Operations would now be directed to the east way of the Maudlin seam where the remaining victims may have been sent by the master shifter.

CHAPTER 18

The Maudlin seam is on fire

Day 24 to 28, 1st to 5th October 1880

Two hundred men were deployed in four teams of fifty men in six hour shifts to remove the debris from the mine. A vast amount of timber was required and miles of tramways had to be entirely re-laid before the pit could be put back into proper working order. The work of the repair teams would go on for many months.

The work of the explorers had been going on relentlessly in the three weeks since the explosion. Not one man in the working parties had tried to lay blame with the men or the officials at the colliery. That was for the Coroner and the jury of the inquest to decide. Instead, day after day the men in the working parties faced the dangers of choke-damp, afterdamp and falling roofs just so that the widows and children would be able to gaze upon the remains of their menfolk once more before they were finally laid to rest in a hallowed burial ground. Paramount in driving the exploration work forward was the Colliery Manager, Thomas Stratton, whose unremitting efforts and labours had won the admiration and grateful thanks of everyone in the colliery village, the mining engineers and viewers from other collieries and the national newspapers.

Search operations had moved from the Maudlin incline which had given up all of its dead to the east way of the Maudlin. The body of Samuel Vickers, horsekeeper, was found at the entrance to the east way. His body was dreadfully mutilated and he was only identified by the trousers he was wearing. An alarming report was brought to bank by the explorers at 10.00 pm on 1st

October. The east way in the Maudlin seam was on fire! The working parties found it impossible to move forward because the way was full of black smoke; scalding hot steam was being emitted; the stone walls had turned red with the heat and roof falls impeded the way. The epicentre of the fire was believed to be at the Maudlin stables. Mr Stratton who was directing the work in the east way sent for the fire appliances which had been used to fight the last fire. All of the officials and workmen with the exception of six men were taken out of the pit because of the fear of another explosion. The fresh current of air directed into the east way to provide better ventilation for the explorers was thought to have fanned smouldering embers in the stables and increased the fury of the fire. Eight fire extinguishers were taken down to fight the fire but there was great concern for the Manager and the six firefighters tackling the inferno. Mr Lishman, Forster and Hall, mining engineers, were urgently sent for at 3.00 am and consulted with the Seaham colliery officials. The executive group decided that the working parties should descend the pit and attempt to re-direct the current of air away from the east way and to organise a plentiful supply of water to fight the fire. Pipes were organised to take water down the shaft and reels of hose were taken down the pit. However, the falls from the roof were becoming frequent and heavy making it very difficult for the men to work. It was feared by the men that the fire was raging so fiercely that it would be impossible to put out. The only feasible solution offered was to starve the fire of oxygen. That could only be achieved if the east way of the Maudlin seam was sealed with a brick wall. If that happened the remaining bodies of their workmates could not be recovered.

The colliery management and the consulting mining engineers consisting of Mr Bell, (Government Inspector), Messrs Hall, Lishman, Forster, Corbett, Stratton, Johnson, Armstrong, Eminson and Turnbull all concluded that the only viable option to save the fire from spreading to the rest of the mine was to seal off the

Maudlin seam. Bricklayers began to build up three stoppings at the main intakes to the Maudlin workings. One of the stoppings was eight feet high, eight feet long and six feet thick and composed of solid brickwork. The other two were slightly smaller. Each stopping was hermetically sealed. A gauge was set to show the state of the heat and two, three-inch pipes with removable plugs to allow examination were inserted into two of the stoppings and a larger pipe was

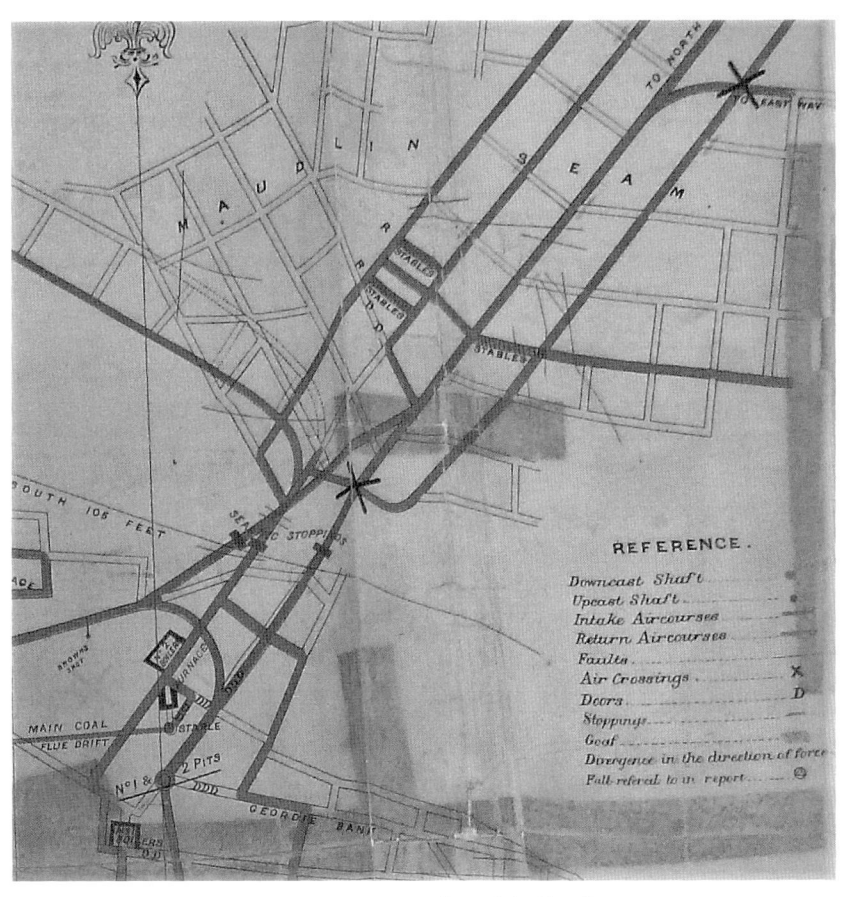

Location of the stoppings sealing off the Maudlin seam
Courtesy of Durham Records Office D/X 1051/57/1/1

inserted into the third stopping by which it could be known when the fire was out. The brick stoppings would remain until it was thought safe to take down the walling. Now there was no way the remaining bodies in the Maudlin could be recovered until it was confirmed that the fire was extinguished. After the last explosion at Seaham Colliery in 1871 a similar stopping was erected under the same circumstances and it was seven weeks before the fire was extinguished and the stoppings were taken down. The work of the exploration and recovery teams was over. More than 1,200 men had been out of employment for almost one month since the explosion. It was now time to put back in order those parts of the pit affected by the explosion so that the men could commence regular employment again. Within two weeks the workmen employed in clearing away the debris and making repairs to the Main Coal, Hutton and Harvey seams had successfully placed the greater portion of the mine back into working order. The Maudlin seam remained sealed up. Despite the hardships felt by the men and their families who had been laid off the general consensus of the majority of the Seaham Colliery miners was that they should refuse to go to work until the bodies of their twenty-eight comrades were recovered. The aversion of miners to continue work after a serious accident to one of their workmates was well-known. There was not a colliery in the district where a single pick would be struck after the hewers had learned that even the smallest putter boy had lost his life. As soon as the news was received every man and boy would make straight for the shaft and ride to bank. This was a long-standing principle that had been upheld for years by the pitmen at Seaham Colliery and other miners in the Durham coalfield.

CHAPTER 19

The adjourned first inquest re-opens

Day 42 to 45, 19th to 22nd October 1880

The adjourned inquest on the bodies of the men killed in the explosion at Seaham Colliery was re-opened at the Londonderry Literary Institute. The Coroner, Crofton Maynard, introduced the legal counsel for the various interested parties to the jury. They were: -

- Barrister, Mr Wright of London on behalf of HM Government
- Barrister, Mr Edge of Seaham Harbour representing the Marquess of Londonderry
- Barrister, Mr Atherley Jones on behalf of the Durham Miners Association

Also present were: -

- Mr Willis and Mr Bell, HM Inspectors of Mines
- Mr Atkinson, Assistant HM Inspector of Mines
- Messrs Foreman, Crawford, Patterson and other officials of the DMA
- Mr JB Eminson, Agent for the Londonderry Collieries
- Mr Corbett, Viewer of Seaham Colliery
- Mr Lishman,
- Mr Hall, Viewer of Haswell Colliery
- Mr Burt, MP for Morpeth

There was only a small attendance of the public. The Coroner and jury examined plans of Seaham Colliery drawn by George Bowdon, surveyor, of Wardley near Gateshead. The plans familiarised the jury with the layout of the mine.

Thomas Henry Marshall Stratton, Colliery Manager described the colliery personnel in the mine and the normal working arrangements. He then explained to the jury how many men and boys were saved and how many were lost in each district on that terrible morning of 8th September 1880.

District	Men and Boys Survived	Killed	Total
No. 1 Hutton Far-off way, 3rd East straight in, 3rd East north side	1	65	66
No. 2 Hutton Brough's way, 3rd West, 1st North, 2nd North, South	43		43
No. 3 Hutton South Way Hutton Seam	4	23	27
Main Coal	19		19
No. 4 Maudlin Incline way, East way		76	76
Total	**67**	**164**	**231**

Of the 231 men and boys that were down the mine at the time of the explosion 67 survived. Of those killed 65 were in No. 1 Hutton, 23 were in No. 3 Hutton and 76 were in the Maudlin. William Laverick, onsetter was the only man who survived in the No. 1 Hutton. The parts of the mine unaffected by the explosion were No. 2 Hutton and the Main Coal seam.

The Manager, Thomas Stratton was questioned at length about the state of the mine immediately prior to the explosion. The jury were informed that the mine was exceptionally dry and dusty. Out of twenty-three collieries tested on a hygrometric scale in 1869 Seaham Colliery was found to be the driest. Mr

Atkinson, Assistant Inspector of Mines agreed that immediately before the explosion the mine was in an especially dry condition. Mr Stratton was questioned about the testing and the presence and reporting of gas and on the ventilation of the mine. Mr Stratton described the number and type of safety lamps and was questioned about the use of open flame lamps. The process of firing shots was discussed and the jury were informed that immediately prior to the explosion it was not unusual for shots to be fired every day. Shotfiring was used to drive stone drifts, take down roofs and form refuge holes. One month before the explosion a miner named Robert Guy was run over by a set of tubs in the Maudlin engine-plane. Remarks made at the inquest on his death prompted Mr Stratton to embark on a project of enlarging those refuge holes that were too small or unsuitable. This work had been ongoing for some weeks.

The inquest then heard evidence from George Carr, Overman of No. 1 pit who described the barometer, water gauge and thermometer readings and his observations on the shift before the explosion. He had visited the east and west stables before going to the far-off way. The master shifter Charles Dawson reported to him that the third east way was alright. The ventilation in all parts of the pit was good. He found no evidence of gas. He came up to bank at 10.00 am on the morning of the 7th September and did not go down again before the explosion. He confirmed that the examiners appointed by the workmen had made their inspection a week before the explosion. He told the jury that he had found very little gas in his district in the past and any that was found were little "blowers" rather than an accumulation of gas in the roof. Pressed by Mr Wright where the explosion came from, he thought from the evidence of the disturbance that it came from the Maudlin side.

John Hutchinson the miner who came out of the pit because of illness a few minutes before the explosion was called before the inquest. He told the jury that he had worked at the pit for seven years and had never heard of any gas being found in it. In answer to a question from Mr Atherley Jones (counsel for the DMA) he hadn't noticed any dust except in the Maudlin seam where it could have been up to an inch thick.

John Miller, overman in the No.2 Harvey seam for five years was last there on the 7th September the day before the explosion. The mine was all right at that time and there was not much gas in his seam.

Robert Barlow, overman at the Maudlin seam was responsible for both parts of the Maudlin. He had examined the east district on the 6th September and the west district on the day before, the 7th September and everything appeared to be in order. There was no indication of gas or danger of any kind. His examinations had not found much gas in the Maudlin in the three months prior to the explosion. He had found a little gas in the west return between the long walls but never any in the east way. Three or four shots were fired the night before the explosion. James Brown had fired a shot that night as it was evident that the debris from the shot had not been cleared away. At the Polka way end he observed that the blast from the explosion had gone both ways.

After further evidence had been heard Mr Atherley Jones on behalf of the DMA asked the Coroner if he could request the Home Office on behalf of the miners at Seaham Colliery to appoint an experimental chemist to explore the theory that coal dust in the workplace may intensify or increase the possibility of an explosion in the pit. The Coroner asked Mr Jones if it was thought by the Association that a shot fired when coal dust was flying about could have been the cause of the explosion. Mr Jones replied that he did not know and that is why they should request a scientific opinion on the matter. He then requested that Messrs Bell and Willis, HM Inspectors of Mines and representatives of the miners go down the mine for the purpose of reporting the presence and quantity of coal dust and to take samples for future examination and analysis.

The inquest was then adjourned until 13th December 1880 to allow time for the fire in the Maudlin seam to burn itself out and be opened up and examined.

CHAPTER 20

The strike begins

Day 59 to 102,   5th November to 18th December 1880

Coal working had gradually resumed at Seaham Colliery by the end of October. Objections by the men to working while the bodies of their comrades in the Maudlin seam were still entombed had apparently disappeared. Additional manpower was sought from other colliery districts to replace the men lost in the explosion and extra ponies were acquired and sent down the pit in place of those killed. In 1871 the Marquess of Londonderry had purchased an estate in Lerwick in the Shetland Isles to breed and rear Shetland ponies to be used in his Durham collieries. Robert Brydon, Chief Land Agent to the Marquess arranged for the stock of ponies to be replenished from the stud farm at Lerwick.

The Secretary of the DMA was informed in a letter from the Home Secretary on 6th November 1880 that he had authorised a Commission to be established to analyse and experiment with coal or other dust found in Seaham Colliery. The Home Secretary requested the colliery management to provide samples of burnt and unburnt dust found in the mine after the explosion. The scope of the experiments and tests were set out by the Home Secretary. They were to ascertain if the burnt dust found in the mine was: -

- Ash or the product of dust burnt by an explosion of firedamp, or

- Whether the remains of the dust had been explosively burnt with or without firedamp, whilst suspended in the air

In addition: -

- How far did the presence of coal dust suspended in the air or lying on the ground or on timbers intensify or extend the area of an explosion of firedamp?

- How far is such coal dust suspended in the air susceptible to explosion by a shot without the presence of gas?

- What were the effects on coal dust suspended in the air or lying on the floor or timbers when a shot is fired in the gallery of a colliery?

The person appointed to lead the scientific experiments on behalf of the Commission was Sir Frederick Abel, formerly the Professor of Chemistry at the Royal Military Academy at Woolwich

At the end of November readings that had been taken through the "relief valve" built into the Maudlin seam stoppings were not satisfactory and gave every indication of fire and intense heat. The workmen at Seaham Colliery had expected the stoppings to be removed before the end of November and the bodies of their twenty-eight workmates to be recovered. The consultative committee of mining engineers at their offices in Newcastle were of the opinion that any attempt to reopen the workings would place an imminent risk to the explorers and would have disastrous consequences to the colliery. Not only that – owing to the readings taken it seemed that there was some leakage in the strata and that additional stoppings were required and should be erected. Over the next week more stoppings were built in front of the three stoppings that had entombed the twenty-eight miners for the previous two months. The colliery management made it known that it was unlikely that the bodies would be recovered until late February or the beginning of March.

This action incensed many of the men who had begrudgingly accepted the return to work. They now felt so aggrieved that they held a meeting in the first week of December. They demanded a stoppage of all work at the pit until the bodies were recovered. A deputation from the men consisting of Messrs Foreman and Patterson of the DMA and Messrs Crozier, Banks and Burt from the local lodge addressed the consultative committee of mining engineers at the Newcastle Offices. After further consideration the committee announced that they were unable to change their decision and that it would be unsafe to open up the Maudlin seam and the stoppings should remain for at least three more months. The men at Seaham Colliery were not happy with this response which made their resolve even stronger that no work should be carried out at the pit until their demands had been met. The unofficial strike at Seaham Colliery was tabled for discussion at the ordinary monthly meeting of the DMA on 18th December. It was made known by the local lodge officials that more than a hundred men were ready to volunteer to take down the stoppings and enter the Maudlin seam to recover the bodies. The bitter and unofficial strike at Seaham Colliery was the main topic for discussion at the monthly meeting of the Durham Miners Association held at their Durham Headquarters on 18th December 1880. A long and earnest discussion took place on whether the Association should support the men now on strike. Some members of the executive expressed surprise that the local lodge should disregard the advice of some of the best mining engineers in the country that it would be unwise and unsafe to remove the stoppings in the Maudlin seam to recover the bodies of their colleagues. Nevertheless, they recognised that the actions of the men were based upon humane and sympathetic principles and should receive their full attention and sympathy considering the miners at Seaham were expected to work in the same pit where their fathers, brothers and workmates were entombed. For those reasons it was resolved that the men on strike at Seaham should receive ten shillings per week from the funds of the DMA during the time they were out on strike. The strike at Seaham Colliery was now official.

Meanwhile, Mr Barrett, the new manager at Seaham Colliery announced to the mechanics and surface men that he had no option but to give them a fortnights notice because everything now stood idle below ground.

The adjourned inquest opened at the Londonderry Literary Institute on 13th December by the Coroner Crofton Maynard. He was informed about the decision not to open up the Maudlin seam for another three months. As a consequence, and because there was no new evidence available to the jury from the exploration of the Maudlin, he announced that the inquest would again be adjourned until 5th January 1881. The inquest would then go over the whole of the evidence regardless of whether the Maudlin seam had been opened or not.

CHAPTER 21

The inquest resumes

Day 120 to 123,   5th to 8th January 1881

The adjourned inquest on the bodies of the unfortunate men killed by the explosion on 8th September opened on the 5th January 1881 at the Londonderry Literary Institute. The same legal counsel and solicitors for the various parties were in attendance in addition to HM Inspectors of Mines and representatives of the colliery management and the DMA. The new manager of Seaham Colliery, Mr Barrett, was also present as were a large number of miners.

A number of witnesses were called to give evidence. Jacob Steel, stoneman, told Atherley Jones counsel for the DMA that he had worked at the pit for twenty years. He had gone down No. 2 pit shaft to the Hutton seam at 10.00 pm on 7th September and was making wagon ways with William Morris and Thomas Taylor about a mile and a quarter from the shaft. At about 2.25am he felt a "swirl of wind" coming from the direction of the pit shaft. They carried on with their work until about 4.00am when they detected a bad smell and a stythe coming away from the shaft. They decided to make their way outbye. On their way to No. 2 shaft they came across some falls with one big fall about fifty yards from No. 1 pit shaft. He made an observation to his mates after seeing the props blown in that "she was blown in from the Maudlin side." There were three or four inches of dust on the tubs. They got forty or fifty yards into the No. 1 Hutton and found the blast had blown everything in-bye. Questioned about the maintenance of the workings he told Atherley Jones that he thought the majority of the men were satisfied with the amount of money expended on safety and the management of the mine.

John Turner, foreman, had worked at the colliery for nine years. His evidence was of paramount importance to the Coroner and jury in arriving at a conclusion about the possible cause of the explosion. He was engaged in making wagon ways in the straight-way in No. 3 Hutton. Before he left the mine on the 7th September, he witnessed Samuel and William Venner, father and son, drilling a hole in preparation to firing a shot at the refuge hole beside the staple. Twenty yards from there Thomas Hindson the Queens Cup winner and his mate Rawlings had already fired a shot. After the explosion John Turner returned to the spot where the shots had been prepared. The stones were still lying where the Venners' had fired the shot but he could not tell whether they had fired it or the explosion had fired it. There were also plenty of stones to be cleared away where Hindson and Rawlings had fired the shot and it was unlikely they would have made another shot whilst the stone from the last shot had not been cleared away.

William Crozier, coal hewer, worked in No. 2 seam east way. On behalf of the local lodge, he carried out inspections of the mine. In July he had examined the No. 3 Hutton seam and found no gas and, on the 6th July, he examined the returns in the east way of the Maudlin. There was a small amount of gas that seemed to come from a small leak in the roof where there had been a fall of stone. The master wasteman arranged to correct it. The ventilation in the mine was pretty fair but better in the Harvey than elsewhere. He considered the system of working the mine was good and he did not know of any false economy. The manager was very strict in enforcing the rules. Only three weeks before the explosion the manager had dismissed Thomas Foster, a deputy who allowed the men to work on the coalface with their lamps about five yards outbye because gas was present. The men heard later that within two weeks of his dismissal Foster had sold up and had sailed to America.

Thomas Burt, hewer, had been working in No. 1 Hutton for more than two years. He had worked in all parts of the mine and was employed by the men to carry out regular examinations. About six o'clock on the day of the explosion he went down the pit with Mr Patterson of the DMA. The indications showed that the explosion had taken place within a certain radius of the staple in No. 3 Hutton between Venners's shot and the staple bottom.

After some observations and further questions Mr Wright, barrister, remarked that if the mine fired in the Polka way it would simply be a question of a defective lamp as there had been no suggestion of a shot having been fired there. Among the many falls of roof which were found after the explosion in various parts of the area affected, one was found in the No. 1 Hutton intake at a point about 550 yards south of the downcast, and a few yards north of the Polka doors where the body of Anthony Ramshaw was found. It had been suggested that if a fall occurred immediately before the explosion and liberated a bag or reservoir of compressed gas this could have been ignited by the faulty lamp.

Mr H Harrison, stoneman, stated that he had been employed at the colliery for forty years. His opinion was that the explosion occurred in the fire holes in No. 1 Hutton. He remarked that the coal dust was very dry in the mine and was collected and sent to bank. He said he did not approve of coal dust being put into refuge holes.

Mr Stratton the former Manager was called to give evidence. He was called to the mine within five minutes of the explosion. Shortly after arriving he went down with Mr Atkinson, Assistant Inspector of Mines. The bottom of No. 3 Hutton was a complete wreck. They found James Brown lying on his back in front of a refuge hole about twenty yards from where it was known he was preparing to fire a shot. The body of his workmate Simpson was found in the

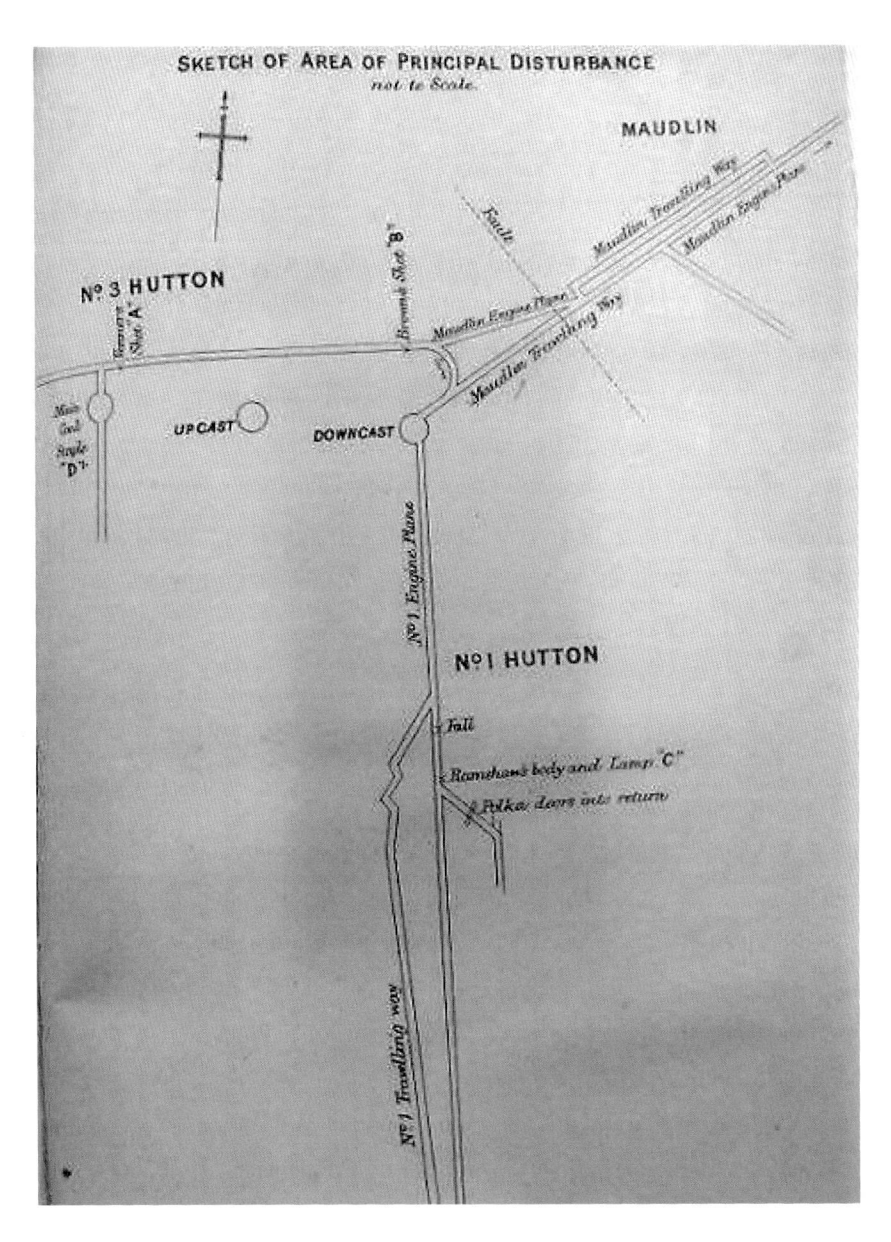

Sketch showing areas of principal disturbance in No.1 Hutton, No.3 Hutton and the Maudlin and the location of Venner's shot (A), Brown's shot (B) and Ramshaw's lamp(C)
Courtesy of Durham County Records Office DRO 1051/5

direction of the No. 1 shaft at the Maudlin engine plane. These would have been the probable places they would have retreated to when firing a shot as they could warn any men coming along the way. At the junction of the travelling way and the engine plane there was a great deal of damage. At the Polka way end they found the doors had been blown inward towards the east return. Going from the Polka way end they found Anthony Ramshaw's body about ten yards from the entrance. He was lying on his back against the pillars. His lamp was found in many pieces near his body. One of his feet was found on the outbye side of him. His tub was about 25 yards away from his body and was completely covered by a fall. At the place where the travelling way joined the main coal staple Samuel Venner and his son had fired a shot.

Mr Stratton answered further questions about the type of safety lamps used in the pit. After further questions Mr Wright, barrister remarked "Is it not rather singular that the particular place in the mine where there was an explosive kind of coal dust was close to the place where Brown's shot was fired?"

Richard Forster, mining engineer at South Hetton and Murton Collieries testified that he was at Seaham Colliery at 8.30 am on the morning of the explosion and together with others he made an exploration of the mine. In his opinion he could trace a direct centre of the explosion from the Polka way. William Fairbairn Hall, mining engineer explored the mine on the Tuesday after the explosion. He agreed with the opinion of Richard Forster that the centre was the Polka way. When questioned about the fire in the Maudlin he stated that it was the unanimous decision of Mr William Armstrong, Mr Boyd, Mr Johnson, Mr Baker Forster, Mr Lindsay Wood, Mr Forster, Mr Lishman, Mr Stratton, Mr Corbett and the government inspector that it was highly dangerous to explore further and that it should be sealed up. He further commented that it would be highly dangerous to open it up at the present time.

After hearing evidence from a number of other witnesses the Coroner indicated that he would have liked to finish the inquiry at that sitting but as Professor Abel's report on the explosive content of coal dust was not ready a further adjournment would be necessary until the 9th February 1881.

CHAPTER 22

Seaham Colliery widows appeal to the Relief Committee

Day 128, 13th January 1881

The Reverend WA Scott, Chairman of the Seaham Colliery Relief Committee received a deputation on 13th January 1881 from unhappy widows of the miners who lost their lives in the explosion. The local press were informed that nearly one hundred widows met with Rev. WA Scott to complain that in the four months since the explosion they had only received four shillings from the funds; those not living in free colliery houses had not received a penny towards their rent and although it had been resolved that the widows would receive three pence per week for each child's school fees nothing had yet been paid. Those children had been sent home by the school for the money. The local newspapers were informed that they demanded to know from the Rev. Scott what had happened to the money that had been subscribed into the Relief Fund.

The Vicar of New Seaham agreed that it was true that more than £12,500 had been subscribed into the fund out of which about one third had been handed over to the Miner's Permanent Relief Fund to help with the very heavy commitment the disaster had placed on the fund. Of the remainder £7,500 had been invested at 4% interest with the River Wear Commissioners and the interest from this and the principal was to be applied to relieve the hardship of the widows and children. The widows claimed that one shilling per week was a small sum and that the widows of the 1871 explosion had received two shillings and nine pence per week. The committee of the Seaham Colliery Relief Fund consisted of fifty members and for some time they could not reach agreement on how the funds should be used. Some of the committee were strongly in favour of donating all of it to the Miner's Permanent Relief Fund which had taken on a heavy liability subsequent to the disaster. Consequently, the committee had found some difficulty in getting any allowance at all to the widows.

The Rev. Scott told the widows that the deputation could lobby or appeal to other members of the committee as they had done with him and he would support their case but he felt that public opinion was against them because of the unreasonable strike by the miners at Seaham Colliery. In frustration at the feeble response to their appeal to the Chairman of the Relief Committee the widows requested the local newspapers to publish the following letter: -

"To the Editor – Will you allow us a small space in your valuable paper to ask your readers and subscribers of the money given for the relief of the widows and orphans of Seaham Colliery explosion if they are willing to let the money be applied as is proposed? We, the widows of Seaham Colliery, think it unfair. There are a great many amongst us that were left with very little, and all we got out of this money is one shilling a week and nothing for the children, only threepence a week to pay their school with, and we think that, if the public knew, they would sympathise with us and try to make it otherwise, for there are many of us in great need of money at present. Signed by THE WIDOWS OF SEAHAM COLLIERY."

Five days later a further letter to the Editor of the Sunderland Echo was published that shone some further light into the monies paid to the widows and orphans. It read: -

"Sir – In your issue of the 14th January you give a report of a meeting of the widows of the sufferers in the late Seaham Colliery explosion. As that report is calculated to mislead the public, it is only fair to give the following statement of

facts: - Each widow has been paid £5 for funeral expenses, besides other small gifts, and is now in receipt of six shillings per week for herself and two shillings per week for each of her children, and has a house rent free except in the case of three or four whose husbands did not have a free house when alive. In addition, threepence is paid for the education of each child of a school age. On average each widow is in receipt of eleven shillings per week and a free house. Yours etc. FAIRPLAY"

One week later a further letter was published in the local newspapers that supported Fairplays' observations. It read: -

"…… Out of the sum of £7,500 invested with the River Wear Commissioners it was resolved by the Relief Committee that this amount would support cases not eligible to relief and also to supplement payments from the Miners Permanent Relief Fund. The widows would receive one shilling per week and if they had to leave their colliery house a further one shilling per week. The school fees of threepence per week would be paid for the education of each child. The balance of the £7,500 after meeting these payments, if any, would be handed over to the Miners Permanent Relief Fund at some point in the future. The widows as you are aware are paid Six shillings per week from the two funds and two shilling per week for each child. The payments to the widows continue during widowhood and to the children until the boys are 12 and the girls are 14 years of age. Were a larger sum paid to the widows for a few years the payments might cease when they stood most in need of the money. I am, yours etc. AN OBSERVER"

A report on the financial health of the Seaham Colliery Relief Fund four years later reported that at 1st January 1885 the amount of money to the credit of the fund was £6,870 of which £6,500 was invested with the River Wear Commissioners and the balance was deposited with Messrs Woods & Co bank of Seaham. The trustees stated that the balance of the funds available justified the original anticipation of the committee and the fund had enough money to carry out the scheme as originally arranged. Ten years later the funds available were just under £5,000 and by 1904 the Seaham Colliery Relief Fund still supported 25 widows and the reports and financial statements showed that the fund stood at £2,097. Five widows were still supported by the fund in 1925 which then stood at less than £1,000.

CHAPTER 23

The strike is causing real deprivation to mining families

Day 133 to 146, 18th to 31st January 1881

During the course of the inquest Mr Stratton informed the jury that after the explosion he had made experiments to ascertain whether there was any danger from the detonation of coal dust in the absence of gas. The results of these experiments were shared with the Institute of Mining Engineers. The conclusion he drew from these experiments was that, in the absence of gas, there was no danger from coal dust exploding even if a shot was fired. He did agree, however, that coal dust was an active ingredient in a situation where there was an explosion. The Coroner had at the suggestion of Mr Atherley Jones requested the Home Office to appoint an experimental chemist to conduct experiments on the part that coal dust may have contributed to explosions in the mine. The Home Office appointed Professor Abel as an independent scientific advisor to carry out these experiments. During January 1881 Professor Abel carried out tests at Gareswood Hall Colliery. In attendance throughout these experiments were a large number of the Royal Commissioners appointed to gather evidence relating to accidents in coal mines. Four of HM Inspectors of Mines were also in attendance together with Mr Atkinson the Assistant Inspector of Mines for Durham. Taking a keen interest during the tests were Mr J Foreman, President of the DMA and Mr WH Patterson, Financial Secretary of the DMA and several viewers from collieries in the district. At the resumption of the adjourned inquest on 9th February the Coroner read out a letter from Professor Abel which indicated that the results of his experiments would not be finalised for some time and suggested that the 12th April would be a suitable date to present his findings. Crofton Maynard, Coroner, agreed that this date should give those officials connected with the mine extra time to get the Maudlin open and so the inquest was adjourned until that date.

A presentation was made one week later to Mr Thomas Stratton by the principal inhabitants of New Seaham and Seaham Colliery to mark his departure from the post of Colliery Manager and his appointment with HM Inspectorate of Mines as a government inspector. Knowing that Mr Stratton had a pronounced objection to testimonials they had agreed that Mr George Turnbull should instead give an address. Mr Stratton thanked those present for the truly exceedingly flattering address and for the beautifully chased silver teapot with inscription presented to Mrs Stratton. He told them he was truly sorry to part with them under circumstances of darkness and disaster but he trusted that brighter days would be granted to them soon.

At the end of January Mr WH Patterson of the DMA visited Seaham Colliery to pay the striking miners their fortnightly allowance. He also intended to take a poll to ascertain whether or not the men would commence work and he found that the feeling of a large number of men was to stay out. It was looking like the stoppage at Seaham would be lengthy and severe. The coal allowance to workmen's colliery houses had been discontinued and the wives and children were suffering severely from the harsh January weather. The allowance paid by the DMA of ten shillings per member and five shillings for half members was quite inadequate to sustain the needs of the families. The DMA had calculated that the cost of supporting the men was £400 per week from their funds which were far from a healthy condition. Many wives and children were suffering hunger and starvation to uphold the principle that their menfolk should not work in the mine whilst the bodies of their fellow workmen were still entombed in the Maudlin seam. Unfortunately, whenever a ballot was taken by the local lodge on whether to return to work or stay out on strike a large number of men who wished to resume work abstained from voting. If they had declared their opinions at the outset the sad and bad-tempered strike which had inflicted so much harm on everyone in the colliery district including shopkeepers and local tradesmen could have been avoided.

CHAPTER 24

Striking miners assault blacklegs

Day 163,   17th Feb 1881

The hardships born by the men out of work with no wages to feed their families and no coal to light their fires for cooking and to heat their homes was too much to bear for some miners. During the previous few weeks some men whose spirit had been broken by the suffering of their wives and children had returned to work.

A meeting of the men on strike had been arranged on 16th February. More than 250 men congregated outside the Union Room at the New Seaham Inn which was far too small for such a large gathering and an adjournment was made to the cricket field. As the men proceeded to the cricket field, they passed through several of the colliery rows where it was known that the men who had returned to work, the blacklegs, lived. In several places their houses were entered by a particular body of strikers and in every case, violence was used against the blacklegs. Five men were viciously assaulted with injuries to their head and face resulting in black eyes and many cuts. The wives of some of the striking miners played their part throwing handfuls of pepper in the faces of the blackleg miners and tin-panning them going to and coming from work. Walking through a crowd of screaming women banging dustbin lids and fire blazers with a poker could not have been a pleasant experience. One man was dragged from his bed and savagely kicked and struck whilst in another house furniture and utensils were thrown into the street. On hearing about the riotous proceedings Mr Corbett and the new manager Mr Barrett went to the scene of the disturbances and pleaded with the men to stop the violence and continue the strike in the same peaceable manner they had previously observed. Mr Corbett had thought that the kindness of Lord Londonderry in allowing the men to remain in their colliery houses whilst out on strike might have prevented such outbreaks of violence. A spokesman for the strikers replied that they had become incensed on hearing that a number of new men were to be brought to the colliery to work and that they had been practically boycotted when they applied for work at other collieries as they were always refused when it was known they came from New Seaham. They insinuated that when their clearances were applied for, they were refused. Mr Corbett replied that he had only been aware of one man who had asked for clearance to work at another colliery and if anyone wanted clearance all they had to do was apply to the colliery office. He was not aware of any new men being brought to the pit but if any did apply and they were acceptable then they would be given work. He reiterated that all of these things were no excuse for their violent behaviour and he appealed to them to go home in a quiet manner. This request fell on unsympathetic ears. The workmen who were to go to work at four o'clock were prevented from doing so by an intimidating mob and men returning from the pit had to be hidden in some colliery outhouses until dark when they were escorted home by policemen. A number of additional policemen were sent for and more were requested the following day. A number of the men who were alleged to have been influential in directing the assaults on the blacklegs were summoned before the Magistrates Court at Seaham Harbour to answer charges of intimidation and assault. The officials of the Durham Miners Association denied they had any knowledge of the intention to assault the blacklegs on that Wednesday. In addition, the officials of the Seaham Lodge stated that the mob acted against their wishes and advice. However, witnesses of the riot testified that the local lodge officials took the trouble to go round the streets with the mob carrying the lodge book in order to point out where the blacklegs lived.

The members of the Consulting Committee of the Coal Owners' Association met at the Coal Trade Offices, Newcastle three days after the riotous proceedings at New Seaham. To the frustration of everyone involved in the strike at Seaham Colliery they announced that "the committee, relying on their experience in other similar cases, is unanimously of the opinion that it would be unwise and hazardous to interfere with the stoppings before the middle of June at the earliest". This was a sad blow to the wives and families of the men on strike. For more than five months the men had done no work or little work and now had to face the prospect of another three months of idleness and no wages. Public opinion was now firmly of the view that the bodies may never be recovered and that the striking men should accept that situation. Under the most favourable circumstances the miners themselves knew that the roof of the Maudlin seam was notably unstable and it was not difficult to imagine what the ways would be like after the seam had been closed for nine months.

CHAPTER 25

Five striking miners charged with assault

Day 167,   21st February 1881

The small court room was filled with colliers and the street outside was crowded with hundreds of miners anxious but unable to gain entrance to hear the case brought against five young men involved in the disturbances at New Seaham the week before. The Chairman of the Magistrates, Rev. Angus Bethune, sat on the Bench with Colonel Allison and Captain Ord to hear the cases against Thomas Morgan, William Aspden, Robert Dunn, Thomas Lannigan and Simeon Vickers. They were all charged with assaulting a "blackleg" William Scott of 31 California Street on the 16th February 1881. The Magistrates heard that William Scott had finished work about noon on the Wednesday and went home and went to bed. A crowd gathered outside his house including the defendants. They broke in and dragged him out of bed into the yard where they severely ill-treated him. Mr Scott said that there were two storeys to his house and he slept in the upper room. He was disturbed by Simeon Vickers dragging him out of bed and down the stepladder into the front room where there was a number of striking miners. Morgan was in the room. He was then dragged into the back room where Lannigan shouted "The ------, bring him out and we will kill him" and he was dragged into the back yard and kicked by Vickers and Morgan. Vickers knocked him down while Aspden held him and he was further kicked by Lannigan and Vickers. He was only wearing his drawers, shirt and stockings. He lost consciousness and when he awoke he was being attended to by Doctor Beatty. His wife testified that she witnessed Dunn, Aspden and Lannigan strike her husband several times. A number of witnesses came forward to speak for the defendants. Mrs Ann Spanton, John Rowse, Robert Brown, Hugh Briney and two other men called McCartney and Wood gave evidence to prove an alibi in respect of the accused men.

Prosecuting attorney Mr HB Wright addressed the court. He pointed out that Scott and all of those other men who had been assaulted or intimidated had a right to go to work if their judgement directed them to that conclusion. In view of the disorderly riots which had occurred the law, he said, must be vindicated and this was a case where exemplary punishment should be inflicted. If a fine was imposed, he questioned whether it would be paid by the defendants themselves and he asked the court that they should be sent to prison. The Bench retired to consider the case and, after a lengthy absence, Vickers was sentenced to two months imprisonment with hard labour. The other men were each sentenced to one month's imprisonment. Simeon Vickers was then charged with assaulting William Shipley. In the same manner as before Vickers and a number of other men rushed into his bedroom and dragged him out of bed pulling his shirt off his back leaving him in a state of nudity. Vickers felled him by a blow on the head with a basin. He was also kicked and beaten. Simeon Vickers was then charged jointly with Jonathan Wylde for an assault on William Harrald in School Street on the same afternoon. Another miner called Roxby was assaulted by Vickers who struck him in the face several times. The defendant Vickers called Thomas Burt and John Bell as witnesses to testify that he had a stand-up fight with Roxby and Roxby was "worsted" in a fair fight.

After retiring, the Chairman gave the decision of the Bench in the three cases. For assaulting Shipley, Vickers was imprisoned for a further two months with hard labour in the House of Correction. For assaulting Harrald he was imprisoned for fourteen days and a similar period for assaulting Roxby thus making five months imprisonment in total. Wylde was sentenced to fourteen days hard labour in the House of Correction.

Michael Hawkins preferred charges of intimidation in connection with the strike against Thomas Friar, George Vickers, Thomas Vickers and Thomas Banks. It was agreed that all of these cases should be adjourned until the next ordinary Petty Sessions day.

CHAPTER 26

Mass meeting of miners from the colliery districts

Day 174 to 182,   28th February to 8th March 1881

By the end of February there were signs that the appetite for strike at Seaham Colliery was coming to an end. Four men went to the colliery offices and agreed to start work. The local lodge committee felt further capitulation could weaken the position of the men on strike and asked the DMA officials at the Durham Headquarters to take a ballot on the question of resuming work hoping that this would conclusively show a majority to remain out on strike. Messrs Crawford and Patterson of the DMA proceeded to New Seaham where they took a ballot which resulted as follows: -

- For resuming work ……… 113
- For continuing to strike …. 82

Mr Crawford and Mr Patterson met with the manager of the colliery, Mr Barrett and Mr Corbett and put to them the terms offered by the men, i.e. that they should resume work on the same footing as they came out. Mr Corbett could not give an immediate answer because the matter was in the hands of the Coal Owners' Association. However, there was a meeting of the Coal Owners' Association on the following Monday which Mr Crawford and Mr Patterson were invited to attend.

The local lodge officials at Seaham Colliery, aware that the resolve of the Seaham men was weakening organised a mass meeting of about 500 miners from Seaham and Ryhope with delegates from Murton, South Hetton, Framwellgate, Hebburn, Silksworth and Rainton. The meeting was held in the

Miner's Hall, Ryhope Colliery and the clear purpose of the gathering was that the strike at Seaham Colliery should continue until the stoppings were removed from the Maudlin seam and the dead bodies removed. The meeting was told that the Coal Owners wanted the men to return to work upon condition that only a certain number of them should be employed. The Chairman of the meeting indicated that he knew what this meant. It meant that the leaders of the Union would be scattered to the four winds of the earth and he hoped that every man in Ryhope and Seaham would stay out until every man had been reinstated. Thomas Burt from the Seaham Lodge moved that the refusal by the mining engineers to agree to the stoppings being removed was due to obstinacy. Miners, he said, may be ignorant concerning scientific questions but he hoped in a few years the miners would be well-informed men. He acknowledged that the mining engineers were men of ability but that from the explosions at Seaham Colliery over the last 28 years the men had come to the conclusion that the stoppings need not remain in. Although he was one of the exploring party after the recent explosion he told the meeting that they were not allowed to go far enough to see if the fire that was supposed to be raging really existed and, therefore "no one could say positively that there was a fire in the Maudlin seam. His motion that the strike should continue until the stoppings were taken out was seconded and carried unanimously. John Sanderson of the Ryhope Lodge then rose to speak. He told the gathering that the manager of Seaham Colliery had invited the men to come back to work but he would not employ the officials of the lodge as they had been a great deal of trouble to him for some time past. He moved that "we the workmen of the county of Durham recommend that unless the manager of Seaham Colliery allows the whole of the men to resume work, as they came out, then the county be laid down. Some would recommend that they should sacrifice members but he did not believe in that." Mr Willis of

Silksworth seconded the motion and added that by sacrificing men they would sacrifice their own interests.

Two days later on 28th February the officials of the DMA met with representatives of the Coal Owners' Association at the County Hotel, Durham to put to them the propositions made by the men at the mass meeting at the Ryhope lodge. The hopes of a speedy resolution at these talks were doomed to disappointment. Reporting back to a meeting of the Seaham miners held in their lodge room at New Seaham Inn it was announced that the Coal Owners' representatives had agreed to allow all the men to start work again with the exception of twenty-six who were named at the meeting but if work was commenced forthwith the cases of twelve of these men would be reconsidered. These terms were received with much disfavour by the men. The meeting was reminded that the notices given to the striking men to quit their colliery houses would expire on the following Monday; the charges of intimidation preferred against several miners who took part in the disturbances a fortnight earlier were due to be heard that Friday and about fifty summonses had been issued for breach of contract.

Fifty-one miners presented themselves at Seaham Magistrates court a few days later on 4th March 1881 to answer charges under the Employers and Workmen Act 1875. They hoped to show why they should not pay two pounds ten shillings each in compensation for breach of contract for leaving their work at Seaham Colliery from 7th to 13th December 1880 without giving the customary fourteen days' notice. In addition, the cases of four miners charged with intimidation adjourned from 21st February were scheduled to be heard. On the cases being called the employer's side requested the Magistrates for an adjournment as they believed an amicable settlement was soon to be reached with the workmen. Mr Atherley Jones representing the men also confirmed that in his opinion the matter would soon be resolved and that there would be no need to trouble the Bench further on the matter. The cases were adjourned until the following Monday. That evening the striking miners met again and after much debate resolved that they would not return to work on any other terms but the reinstatement of every man including the union officials.

CHAPTER 27

The strike is resolved

Day 185 to 248,   11th March to 13th May 1881

A notice was posted at Seaham Colliery intimating that the striking miners who occupied colliery houses and had received notice to quit but had not done so by Tuesday 15th March 1881 would be evicted. Urgent talks were held with the striking men in the New Seaham Inn. Mr Patterson of the DMA and Mr Corbett on behalf of Lord Londonderry put forward a suggestion to resolve the strike and this was accepted by those at the meeting. Details of these proposals were taken by the DMA the following day to a meeting quickly convened with the Coal Owners' Association at the County Hotel, Durham. These proposals were agreed to by the Coal Owners' Association. The terms of resuming work now agreed between the DMA and the Coal Owners were laid before the men at the New Seaham Inn. The men reluctantly had to face the fact that the Coal Owners would not waver in their resolve and agreed that those union men objected to by the owners should be "sacrificed" and receive one month's notice to leave. The men began to assemble at the colliery offices on Monday evening, the day before the eviction notices were due to expire, for the purpose of being re-engaged. Work was resumed on 15th March 1881. The strike was over. It was reported that the explosion and subsequent strike at Seaham Colliery had cost the Marquess of Londonderry almost £18,000.

Preparations had already been made by the colliery management to carry out the evictions on the 15th March. However, because the strike was over the evictions would not now be carried out. Anticipating many homeless families from the evictions the drill shed of the 2nd Durham (Seaham) Artillery Volunteers on

Station Road had been fitted out with wooden bedsteads and cooking apparatus. A large number of policemen from various parts of the county and about 200 "candymen" arrived at Seaham Colliery on 15th March to carry out the evictions. They were not needed and were sent home. Meanwhile at Seaham Magistrates Court the cases of 51 men charged with a breach of the Employers and Workmen Act were withdrawn. Their cases were dismissed. However, the cases of intimidation were heard. John Coil and George Vickers pleaded guilty to intimidating George Boyd on 16th February and were found guilty and fined £10 each with costs. Several other men similarly charged pleaded guilty and were bound over to keep the peace for six months. The charge of intimidation against John Bell, president of the Seaham Lodge was withdrawn.

Peace at Seaham Colliery lasted only one month. The owners had agreed to re-employ all the men who were out on strike with the exception of eight who had been found guilty of intimidation or were marked men because they were the union leaders and regarded as the instigators of the strike. The names of the men that the employers refused to re-engage, the "sacrificed men", were: -

- Thomas Banks (Committee member)
- John Bell (Secretary of the local lodge)
- Robert Brown (Committee member)
- Thomas Burt (President of the lodge)
- David Corkhill
- John Furness
- Thomas Neasham (Union delegate)
- Ralph Pallister

The applications for reinstatement from these eight men were held over for one month. That month had elapsed and the owners declined to employ them. Consequently, the whole of the Union men employed at the colliery brought their gear to bank and the strike was recommenced. Two members of the DMA executive were sent to Seaham to endeavour to find a resolution to the dispute. The men returned to work whilst the DMA arranged a ballot of miners employed at the Londonderry Collieries at Rainton and Silksworth as well as at Seaham. The results of the initial ballot were conflicting. Seaham and Silksworth collieries were in favour of a strike but the Rainton collieries were against a strike. Comparatively few of the men had attended the union meetings at which the above ballots were taken. In order that a more accurate opinion of the men could be assessed two members of the DMA executive delivered a ballot paper to every union member employed in the Londonderry collieries with only two questions. "Strike or no strike". Mr Patterson of the DMA and Mr Bell the local secretary went around each house at Seaham Colliery and the other collieries with the ballot bag to collect the papers. The papers were taken to the Durham offices of the DMA and counted. The results were: -

	For Strike	No Strike
Adventure (Rainton)	101	119
Letch (Rainton)	69	131
Silksworth	169	493
Seaham	203	100
Total	542	843
Majority against a strike		301

Seaham Colliery was the only one of the four collieries where the majority favoured a strike. A meeting of the men on strike was held to consider the resolution that the men should go to work and that the men objected to by the owners should receive one month's notice to leave. This resolution was agreed to by the men. The miners at Seaham Colliery conceded that they had to accept the mutual right of the employer to engage whom they liked just as it was the right of the employee to work for whom they chose.

CHAPTER 28

The fate of the "sacrificed men"

Day 233,   28th April 1881

The Seaham Weekly News published a letter from Mr VW Corbett to the Editor of the newspaper in its edition of 29th April 1881. It read: -

"Sir, I am given to understand that the miners in this county have got the impression that the coal owners are aiming a blow at the Miner's Union by the dismissal of some of the Union leaders from Seaham Colliery. Allow me to state most emphatically that there is no such intention. The eight men who were under an arrangement to leave the Colliery were not picked out because they were officials of the Miner's Union but for the reason already stated by me in the press, and as a matter of fact the Miner's Union do not tell me who are their officials. The agreement entered into on 12th March stipulated that the eight men should leave the colliery in one month and give up their colliery houses by that date. I have therefore begun the unpleasant proceedings of evicting them from the houses they occupy."

One of the eight men had already moved to Seaham Harbour and three others had left the colliery by the time this letter appeared in the local newspaper. This left four men, Messrs Banks, Brown, Burt and Neasham who had not given up possession of their colliery houses. They had been allowed to remain in the occupation of their houses until 28th April and they had openly stated that they would not leave. Sixty policemen arrived at the drill shed of the 2nd Durham (Seaham) Artillery Volunteers under the supervision of Superintendent Scott of Castle Eden and Inspectors Webster, Avrill and Smith.

Mr VW Corbett, Colliery Viewer and Agent
Courtesy of Seaham Family History Group

They waited until the appointed time for the evictions. The ordinary "candymen" usually employed in such evictions were not brought to the colliery. Instead, Messrs Turnbull and Brough, under viewers, and Mr Wilkinson Rowell, engineer, accompanied by a body of constables and four foreman mechanics from the colliery acted as bailiffs. They proceeded to the home of Thomas Burt in Butcher's Row. The house was emptied of its contents, furniture and household goods which were placed carefully on the road outside the house. There was very little animosity displayed by the assembled crowd. Mr Burt did not come out of the house until the final articles were removed. The door and windows were fastened up and the eviction team then moved on to the house of Robert Brown in the same street. There a similar scene took place. Next the "bailiffs" moved to the house of Thomas Banks in Mount Pleasant and then on to the house of Thomas Neasham in Australia Street. Although the crowd was much larger in Australia Street there was no abusive language or ill-behaviour. In all cases the furniture was carefully placed at the back door and everything was done peaceably and quietly. The weather was fine and the furniture did not suffer any damage. Mr Corbett, Colliery Agent had given instructions for the conveyance of the furniture of the evicted men on colliery carts to any place they wished to go. All four of the evicted families obtained houses at Seaham Harbour and their furniture was moved there shortly afterwards. Mr Corbett sent men from the colliery to assist in the packing, loading and unpacking of their furniture at their new houses.

It was clear that these eight miners were marked men and would never get employment in any of the Londonderry or the Earl of Durham collieries. In fact, it was unlikely that they would ever gain employment in the coal mining industry again. The DMA granted each of these men £50 to start a new life. An adventure across the Atlantic Ocean to the United States of America was chosen by most of them. Two years earlier a miner's conference had been organised in

Manchester to consider a scheme to assist prospective emigrants from British coal mines. Mr Crawford, President of the DMA took a trip to America in one of the Inman liners and wrote an account of it entitled "In the Steerage" for the union members. A report had circulated in the press describing the foul conditions of the accommodation provided to third class passengers. Mr Crawford went to New York in third class accommodation and completely exposed the untruthfulness of it easing the minds of many miners preparing to leave the country. For almost two decades a steady stream of coal miners had left Britain to begin a new life in America. Peter Lee who was later to become General Secretary of the DMA and then President of the Miners Federation of Great Britain immigrated to the USA in 1886 and worked in the coalmines of Ohio, Pennsylvania and Kentucky for two years before returning to Wingate, County Durham. John Wilson one of the founder members of the DMA immigrated to the USA in 1864 where he worked in the mines of Pennsylvania and Illinois returning in 1867.

Three months after the evictions on 25th July 1881 a party of emigrants left Seaham for Liverpool bound for America. Six of the eight men who had been refused employment i.e., Messrs Burt, Brown, Banks, Neasham, Bell and Pallister and their families prepared themselves for a different and hopefully better life in the new world. There is no record of John Furness or David Corkhill in the USA passenger lists although the family records of David Corkhill record that he went to the USA and returned to Seaham and became Colliery Checkweighman in 1885 and a parish councillor in 1896. As Colliery Checkweighman the colliery management had no authority over him as he was employed by the workmen. The six families immigrating to the USA left Seaham by train at 9.00am on 25th July to hearty cheers from a crowded platform. At Sunderland they were greeted by numerous friends and on arriving

AMERICAN LINE

Report or Manifest of all the Passengers taken on board the Sh. British Queen whereof Leonard Nowell is Master, from Liverpool via Queenstown (burthen Tons 2277) and owned by The British Ship Owners Co Limited at Liverpool and bound to Philadelphia

NAMES	AGE	SEX	OCCUPATION	To What Country belonging.	Country of which is their intention to become Inhabitants.	Number and Names of Passengers who have died on the Voyage.
Thomas Hurst	29	M	Labourer	England	United States of America	not
Dorothy Hurst	28	F	Wife	"	"	"
Amelia Hurst	2	F	"	"	"	"
Mary Hurst	1	F	"	"	"	"
John Thos Hurst	4/12	M	"	"	"	"
Thomas Mallan	40	M	Labourer	"	"	"
Matilda Mowell	38	F	Wife	"	"	"
Herbert Stacey Nowell	7	M	"	"	"	"
Alfred Nowell	25	M	Labourer	"	"	"
Alfred Lego	24	M	"	"	"	"
John McC.	28	M	"	"	"	"
William Bell	32	M	"	"	"	"
William Newman	11	M	"	"	"	"
James Newman	40	M	"	"	"	"
Thomas Needham	57	M	"	"	"	"
Jane Needham	57	F	Wife	"	"	"

16

Manifest of the British Queen presented to the Department of Immigration, Philadelphia, Pennsylvania

AMERICAN LINE.

Report or Manifest of all the Passengers taken on board the *Sh British Queen* whereof *Samuel Powell* is Master, from *Liverpool via Leventon* burthen *Tons 2277* and owned by *The British Ship American Colone* of *Liverpool* and bound to *Philadelphia*

NAMES	AGE	SEX	OCCUPATION	To what Country belonging	Country of which is their intention to become Inhabitants	Number and Names of Passengers who have died on the Voyage
Jane Smith Hawthorn	18	F	Spinster	England	United States of America	nil
John Bell	35	M	Labourer	"	"	"
Jane Bell	31	F	wife	"	"	"
Jane Bell	10	"	"	"	"	"
George Bell	8	M	"	"	"	"
Robert Bell	6	"	"	"	"	"
John Bell	4	"	"	"	"	"
Annie Bell	3	F	"	"	"	"
Elizabeth Bell	inft	"	"	"	"	"
James Ford	26	M	Labourer	"	"	"
Robert Annin	49	"	"	"	"	"
Thomas Beaste	63	"	"	"	"	"
Mary Ann Beaste	24	F	wife	"	"	"
Mary Elizabeth Beaste	10	"	"	"	"	"
Ralph Collister	40	M	Labourer	"	"	"
Robert Dodds	27	"	"	"	"	"

Manifest of the British Queen presented to the Department of Immigration, Philadelphia, Pennsylvania at York they got out of the train for refreshments. They arrived at Liverpool at 7.00pm and were put up for the night in a temperance hotel. Next morning, they boarded the British Queen and sailed out of Liverpool at noon. The British Queen was a 2,277-ton steamer, 410 feet long by 39 feet breadth built in Belfast and owned by the British Ship Owners Company. Two passengers who travelled with the exiled Durham miners and their families described the journey to America.

"There was much hustle and bustle on board, everyone being eager to see to the safety of their own luggage. We had not been long under steam when dinner was served and we found the grub satisfactory then, and during, the voyage. At Queenstown many more passengers joined and more provisions were taken on board. On the 27th July the British Queen sailed out for the Atlantic Ocean.

It was not long before many of the passengers complained of sea sickness and some hardly recovered during the voyage. After twelve days sailing, we landed in the Quaker city of Philadelphia. There was a mighty cheer when the pilot came on board. This was a most convenient place for immigrants to land for the trains come right to the boat landing. There is no need to go into the town for at the depot you can be accommodated with everything that is needed – refreshments, rooms, places to have a bath and an exchange for money. After an hour stay we saw our luggage on the baggage cars and got on board a west bound train. We had no trouble to seek railroad tickets as we had our pass from Seaham right through to Alton City on the Mississippi about 25 miles north of St Louis. The trip out West was tedious when you have to ride three or four days and nights. Our numbers from Seaham were gradually growing smaller as parties were dropping off at the different stations as they were nearing their place of destination. The last of our Seaham party left us at Indianapolis when we had many more weary miles before we reached Alton."

The sacrificed men and their families did not appear to act in the way that beaten men would behave. They seemed to have plenty of spirit even after undergoing an uncomfortable and long sea journey over the Atlantic Ocean. Upon their arrival in America the exiled Seaham miners wrote a letter addressed to the owners of the "British Queen" steamship company. It read: -

"Gentlemen, We the undersigned, being as passengers in your steamship called the British Queen do declare and deem it our bounden duty to give our best thanks for the care, safety and comfort that you have bestowed upon us during our voyage from Liverpool to Philadelphia, which one and all can testify in the following terms. 1st Food – We really believe that you have bought the best articles in the best of markets for the comfort of emigrants; we can give our thanks for the abundance of it to suit all, and also that it was cooked clean and pure to the satisfaction of us all. 2nd Civility – Your servants have done all and everything in a good and business-like way, and have treated us with all the civility that can be bestowed upon man, and have acted in a free and business manner, so that we all felt quite at home with them, and we feel great pleasure in giving our best thanks to them for their energy and trouble for us. 3rd Cleanliness – We also deem it our duty to report that everything was kept clean, healthy, comfortable and in good order. Therefore, we can confidently recommend any persons to emigrate in your vessel and under the supervision of your kind and trustworthy servants. We remain, dear Sirs, Signed by Messrs Burt, Bell, Neasham and other miners late of Seaham Colliery and on behalf of all steerage passengers"

Pennsylvania was the heart of the anthracite and coal producing area of America and was the popular destination of many British coalminers. One can only wonder if the American Coal Owners realised what they were letting themselves in for when they hired the "sacrificed men" from Seaham in their mines. The American coal industry had the greatest need for labour organisation because of the appalling rates of injury and death. British immigrants with their tradition of agitating for better and safer conditions were natural trade union organisers. The coal owners hired British miners for their ability to exploit seams that others could not but they did not realise that many men like Thomas Burt, Thomas Banks, John Bell and others also had vast experience of unionism to share with their fellow miners in America. Many British miners stood side by side with their American workmates to establish trade unions, co-operatives and friendly societies.

In addition to David Corkhill who returned to Seaham Colliery to become Check Weighman only one of the other six exiled families who made the momentous leap into the unknown in 1881 is known to have returned to Britain. Thomas Burt returned to England with his family about four years later but not to Seaham and the colliery districts of Durham. He returned to Northumberland where he obtained work presumably with the help and influence of his cousin Thomas Burt, MP for Morpeth. A daughter Elizabeth had been born in America on 15th November 1882 and then a son Thomas was born in Bothal, Northumberland five years later in 1887. The Burt family were living in Tenth Row, Bothal Demesne near Ashington colliery at the time of the 1891 census. By 1896 Thomas Burt seems to have become disillusioned with the life of a collier and the struggles of trade unionism and he applied for the position of Collector of Rates for Ashington Urban District Council. He was unsuccessful in his application and was still recorded as a miner living at Front Street, Newbiggin-by-the Sea in the 1901 census. Thomas Burt was an "edge dweller". There was no walking down the straight easy road for Thomas. He always lived on the edge; stood up for his principles; never dodged a battle and was prepared to fight for the mining community at any cost. He was part of the exploration team that dug through roof falls, debris and dead bodies in the search for survivors of the explosion. He fought to uphold the long-held mining tradition that miners came out of the pit after a fatal accident until the body was brought to bank. He was prepared to suffer cold, hunger and deprivation with his fellow union men in a long strike for those principles. This led him to be sacrificed along with seven others, in the agreement between masters and men and not only lose his house and his job but to also face the prospect of never being employed in the Durham coalfield again. He accepted exile for himself and his family in order to find a better life and travelled to Indiana in America in search of work. The family settled in Coal Creek, Indiana so named because of the large deposits of coal along its banks and where mines extracted large quantities of bituminous coal. One of the attractions of Coal Creek may have been the well-established Methodist Church built twenty years earlier. The Burt family had strong family ties with Methodism. His Uncle, Andrew Burt, in 1877 was a founder member and laid the foundation stone of New Seaham Independent Methodist Church in Enfield Road, New Seaham. After four years working in Coal Creek, Indiana he returned home and ended up back where he started – in the coal mine at Ashington, Northumberland. After a full, frantic and troubled life he died aged 51 at Newbiggin by the Sea in 1904.

CHAPTER 29

The verdict of the first inquest is declared

Day 217 to 219,   12th to 14th April 1881

The inquest to find the cause of the Seaham Colliery disaster resumed on the 12th April 1881. The Coroner, Crofton Maynard asked each of the legal counsel present if they intended to call any more workmen to speak or give evidence on the condition of the pit at the time of the explosion. Mr Atherley Jones representing the DMA told the Coroner that he had made it known that any miner who thought they could provide useful evidence should come forward but none had come forward. The proceedings would therefore focus on the scientific report of Professor Frederick Abel, President of the Institute of Chemistry. He had submitted copies of his report outlining the results of his experiments to the jury and the legal counsel present. The Coroner directed the attention of the jury to the summary given at the end. The results of the experiments were summarised in the report as follows: -

a) That coal dust in mines promotes and extends explosions in mines when it is suspended in air currents and

b) That coal dust itself can become a fiercely-burning agent which will carry flame rapidly when mixed with a proportion of firedamp even as small as 2½ percent. On its own that amount of firedamp would not itself present any danger.

c) That coal dust can create an active explosive effect in a mixture of small proportions of firedamp and air

d) That dust in Seaham Colliery can be a source of danger even if it contains only a small proportion of coal or combustible matter.

In conclusion Professor Abel told the jury that a large volume of flame such as the disturbing effect of a blown out shot igniting coal dust together with very small quantities of firedamp could give rise to explosions in coal mines.

Upon examination Professor Abel agreed that coal dust on its own is not explosive in the absence of gas mixed with air. Mr Wright, barrister asked Professor Abel whether it would be a proper thing to water or remove the dust to which Professor Abel agreed.

Professor Abel had visited the mine as far as Venner's shot towards the No. 3 pit and as far as the Polka doors in No. 1 pit. When asked whether the explosion could have been caused by Brown's shot his opinion was that Brown's shot was not a blown-out shot but it might have been an over-charged shot and there could have been some considerable volume of flame propelled by it into the workings. There certainly were large amounts of dust deposits in a fine, flour-like state in the immediate vicinity.

Mr John Foreman, President of the DMA said he had gone down the mine and looked at the evidence of the explosion. He told the jury that he had come to the conclusion that the shots by the two Venners and by Brown and Simpson were fired simultaneously and that these shots contributed to the explosion. The two men, in both cases, were found lying on the ground about the distance they would have gone apart waiting for the shot to fire and to warn persons coming in either direction. Mr Stratton stated that the shots went off simultaneously with the explosion and witnesses had confirmed this opinion. There was a large amount of coal dust in the pit before the explosion and witnesses measured it to

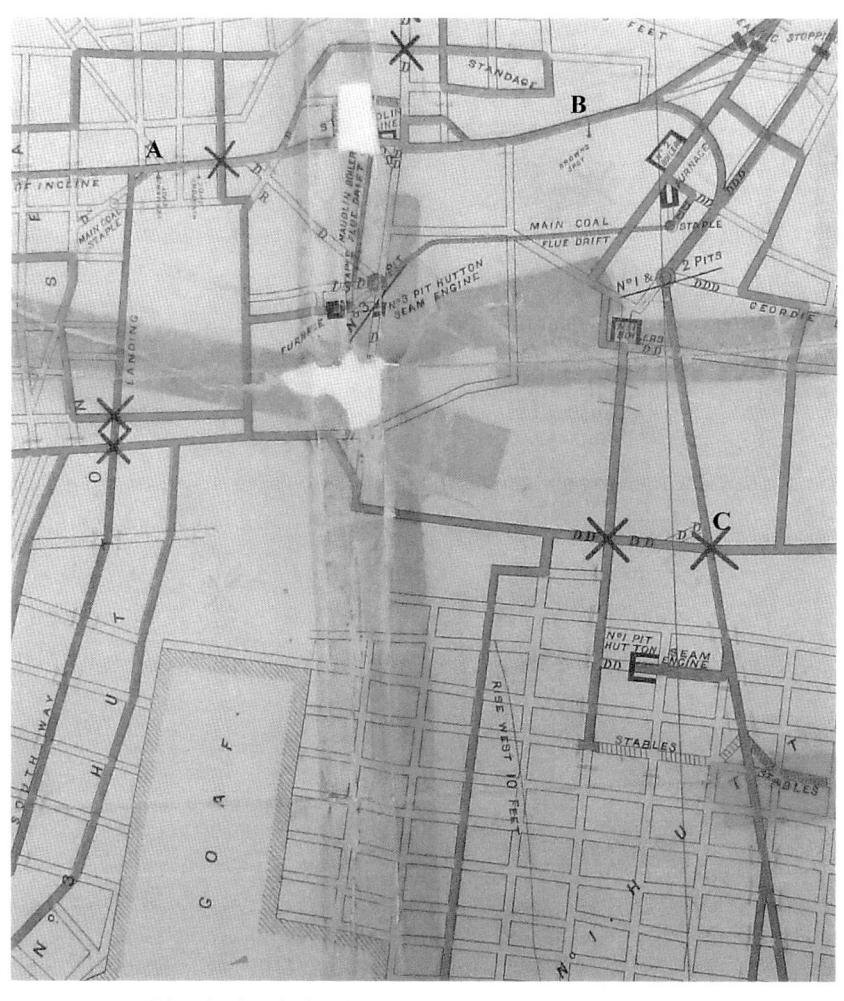

*Plan showing the location of Venner's shot (A), Brown's shot (B)
and Ramshaw's lamp(C)
Courtesy of Durham Records Office D/X 1051/57/1/1*

a depth of six to eight inches. Witnesses attested that the dust was all around the Maudlin curve and that it must have been accumulating for years. Mr Atkinson, assistant Government Inspector of Mines stated that he was one of the first to go down the mine after the explosion. In his opinion the explosion began between the curve and the Maudlin engine and he thought it was Brown's shot and not Venner's that had fired a combination of dust with air and firedamp.

Witnesses examined previously had also put forward a "roof fall" theory to the cause of the ignition that fired the explosion. Among the many falls of roof discovered after the explosion was one found in the Hutton intake about 550 yards south of the downcast in the area of the Polka doors. This fall of stone may have liberated a reservoir of gas which may have come into contact with Ramshaw's faulty lamp blowing it to pieces and igniting the explosion.

In essence the jury were faced with two possible causes of the explosion. They were: -

- a fall of stone releasing gas which ignited on Ramshaw's faulty lamp at the Polka way or,

- A shot fired by the two Venners at the main coal staple or a shot fired by Brown and Simpson at the curve between the two shafts.

Witnesses supporting the Ramshaw lamp theory included all of the workmen and officials called by the colliery management.

Witnesses supporting the shot fired by Brown included the Government Inspectors of Mines Messrs Bell and Willis and also the miners' representatives and the DMA.

The one neutral opinion belonged to Mr RS Wright, barrister and Chief Government representative with no vested interest on behalf of the coal owners or the miners. His opinion was that the weight of evidence was in favour of the view that the explosion started at Brown's shot and gathered strength from the clouds of dust raised in its way along the curve. He also remarked on the similarity of the explosion at Seaham Colliery in 1871 which was said to have been coincident with the firing of a shot and at the same spot. In his official report to the government, he stressed the dangers of coal dust and inadequate safety flame lamps to colliery management. The indications and signs of damage and disturbance throughout the colliery were illustrated by W.N & J.B Atkinson, Mines Inspectors some years later.

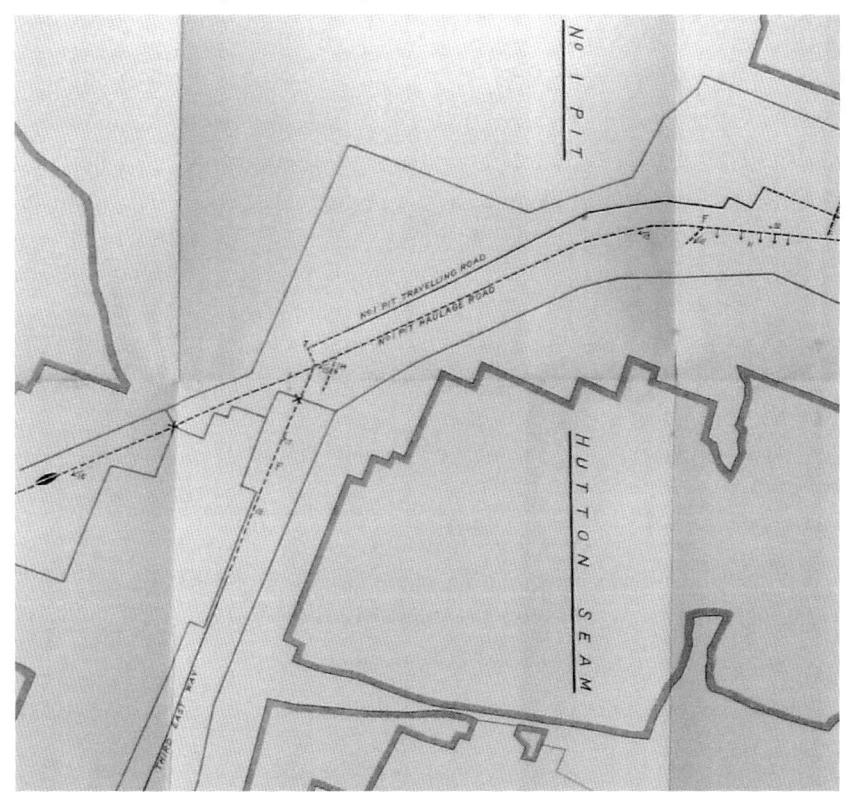

Map showing roadways damaged by explosion and fire (See key to drawing below)
Courtesy of Brian Scollen

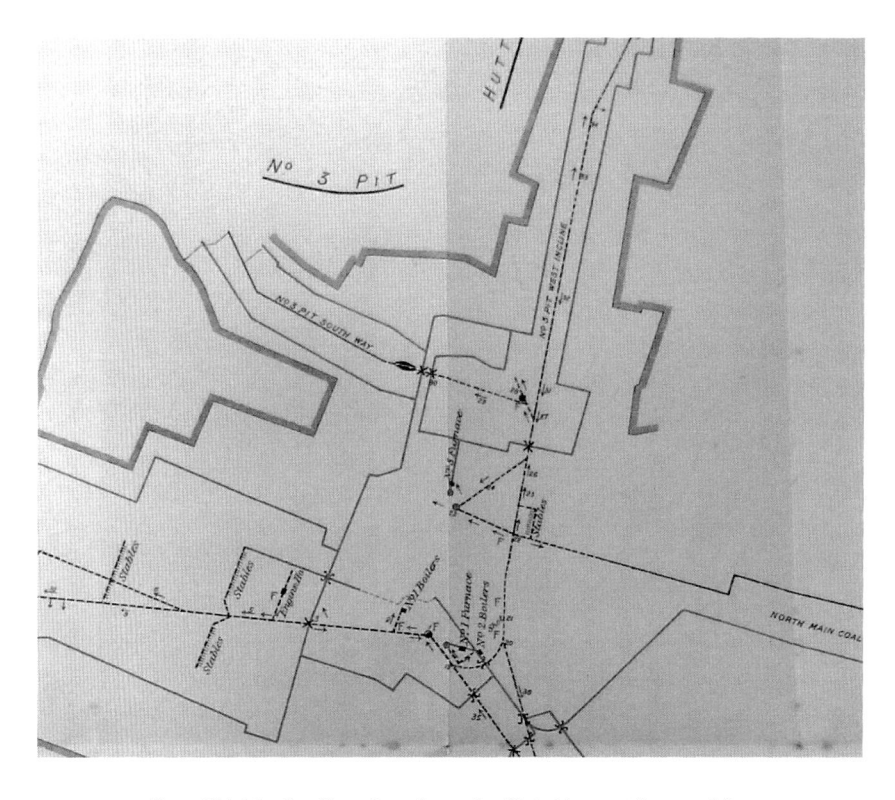

Key: Thick broken lines show the roads affected by great force and fire
Thin broken lines show the roads effected by fire but little force
Unbroken lines show roads not affected by fire or explosion
F show places where signs of fire were evident

Courtesy of Brian Scollen

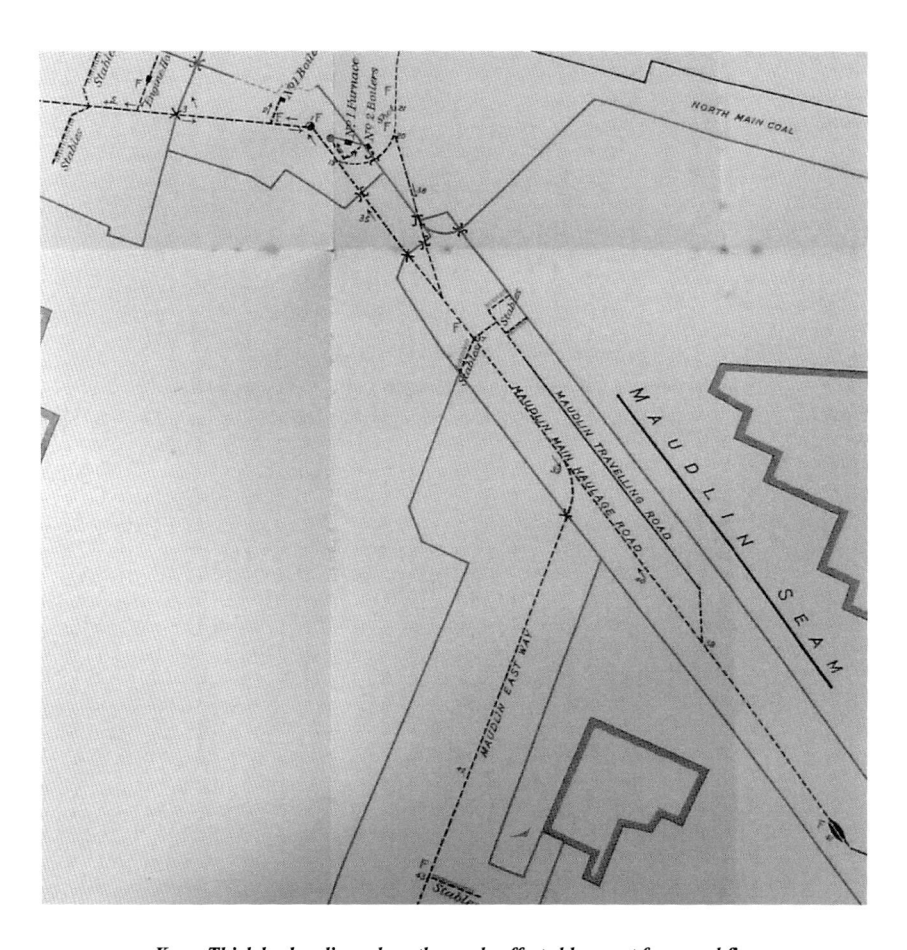

Key: *Thick broken lines show the roads affected by great force and fire*
Thin broken lines show the roads effected by fire but little force
Unbroken lines show roads not affected by fire or explosion
F show places where signs of fire were evident

Courtesy of Brian Scollen

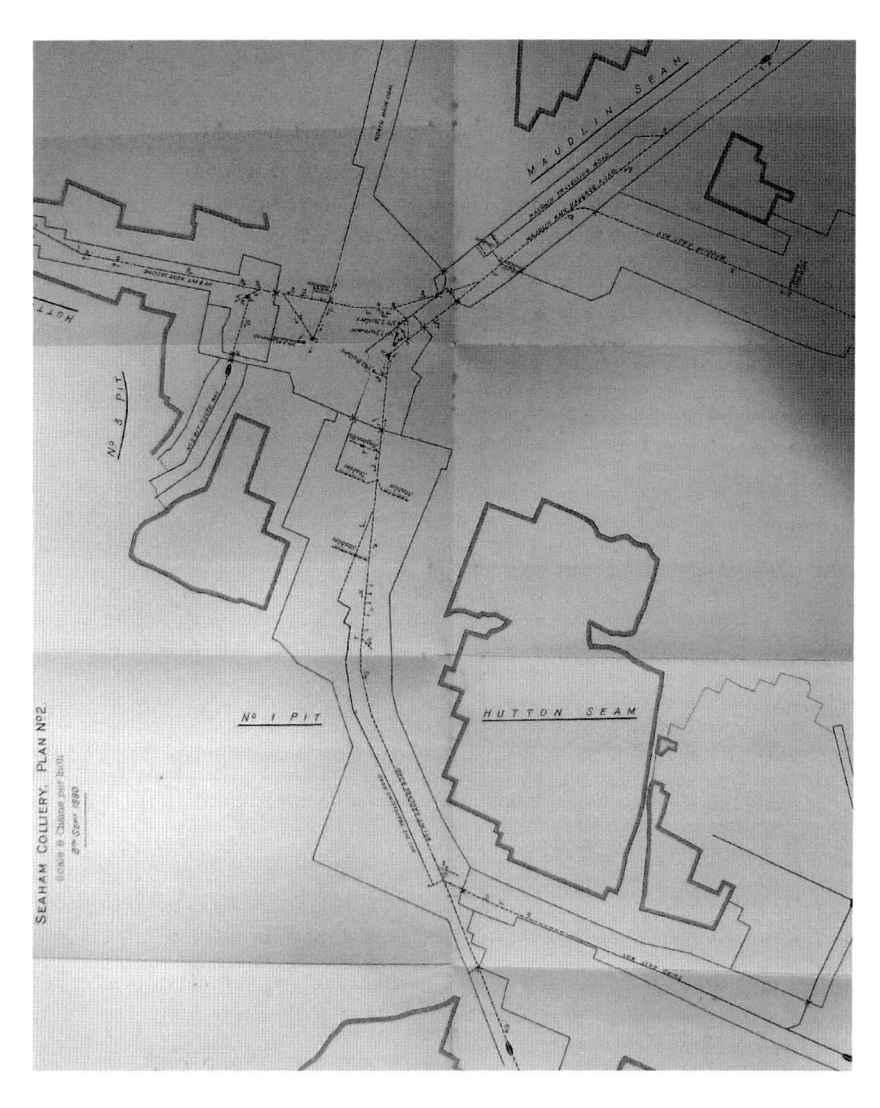

Key: *Thick broken lines show the roads affected by great force and fire*
Thin broken lines show the roads effected by fire but little force
Unbroken lines show roads not affected by fire or explosion
F show places where signs of fire were evident

Courtesy of Brian Scollen

Mr Bell and Mr Willis, Inspector of Mines gave a synopsis of their report to the Coroner. They had found that the whole field of the explosion was largely to the intakes with very little damage being in the returns. The main ventilating doors were all blown inwards in the vicinity of the shafts from the intakes to the returns. The main crossing also appeared to have been destroyed by the force from the intakes. The Inspectors observed that the explosion took place outside of the limits of the Maudlin seam workings. Professor Abel's report had shown that as little as 2.5% of gas in a current of air combined with Seaham Colliery dust was sufficient when passing a lamp flame to become ignited and produce explosive effects. The Inspector of Mines pointed to the likelihood that Brown's marrow Simpson would have kicked up coal dust when he was retreating to take cover from the place of the shot and the fine dust would have been suspended in the airway. In addition, they pointed out the heavy contamination of the air between the curve and the engine of the Maudlin. Bearing in mind the indications were that Brown's shot was an overcharged shot that was likely to produce a flame the situation brought all of the elements together necessary for an explosion. The Inspectors report to the Coroner concluded that the source of the explosion originated at either the curve or the staple and that the curve was the most likely location.

There was no further evidence brought before the inquest and the Coroner addressed the jury. He reminded them that this inquest was not to decide upon a criminal charge or a charge of negligence on the part of the mine owners. No carelessness had been suggested at any time during the proceedings and many witnesses, workmen and officials had certified under oath that everything had been done to keep the pit in a proper and safe condition. He then put the following questions to the jury for their consideration and to assist him into reaching a verdict: -

- Did the evidence satisfy them that the explosion happened at the Polka way end, or through a sudden outburst of gas or,

- Were they satisfied that it occurred through shotfiring?

If they replied in the negative to the first question and in the affirmative to the next question,

- Was the shotfiring at present carried out in a safe manner and whether further safeguards and restrictions should be made?

- Taking into account Professor Abel's report that coal dust was one of the principal causes of the explosion was it necessary for the safety of men in mines that such coal dust should either be watered or removed?

The jury then retired. After half an hour the foreman of the jury returned to the main hall and informed the Coroner that they were unable to determine the seat of the explosion and they therefore returned an open verdict. In addition, they felt that the firing of shots and clearing away of dust were decisions that should be taken by the management. An "open verdict" is normally arrived at where there is not enough evidence available to a jury to return a verdict. This is a rare occurrence and is normally only used as a last resort. It is understandable that the jury may not have been able to come to an opinion about the cause of the explosion as there were three distinct theories put forward each one supported with good arguments. However, to abstain from giving an opinion on additional safeguards when firing gunpowder in the mine or eliminating potentially explosive dust by removing it from the mine is baffling and inexplicable.

Mr Atherley Jones, legal counsel for the DMA reminded the Coroner that when the Maudlin seam was reopened there would have to be another inquest. There was still a great deal of discontent amongst the men that the stoppings in the

Maudlin seam were still in place and he asked the legal counsel representing Lord Londonderry if the men could have an assurance that it would be opened with the least possible delay so that the bodies could be taken out and given a decent burial. Mr Edge for Lord Londonderry replied that this was in the hands of the Board of Management comprised of eminent mining engineers. The Coroner responded to Mr Atherley Jones that he could not take the responsibility upon himself of ordering the seam to be opened. The Coroner thanked the legal counsel, government inspectors and the jury for their patience and then closed the inquest.

CHAPTER 30

The Maudlin seam is reopened

Day 256 to 332,   21st May to 5th August 1881

A meeting of mining engineers was held on Saturday 21st May 1881 at the Wood memorial Hall, Newcastle to consider the question of opening the Maudlin seam. It was decided to defer any decision for a fortnight. Two weeks later the mining engineers met at Seaham Colliery and after a minute examination of the stoppings it was decided to begin operations to take the stoppings down commencing on 25th June. It was believed that the falls in that part of the pit where the remaining dead bodies lay were very heavy and numerous.

Everyone in the colliery village was full of anticipation in the weeks and days leading up to 25th June. For at least twenty-eight families the wait had been too long. They were to find, however, that it was not as simple as taking down the brick stoppings and entering that crypt. An important series of experiments had to be carried out first. No-one knew what to expect when the stoppings came down. Smoke, heat, afterdamp, no air or ventilation, water, roof falls or any combination of such hazards may have to be encountered. The mining engineers were aware of a new breathing apparatus that had recently been patented and they acquired a number of these to experiment with and use in the Maudlin seam if they proved satisfactory. The inventor Fleuss had developed a "patent noxious gas apparatus" which it was claimed would allow the wearer to proceed into the passages of a mine full of noxious gases without injury or danger. It was a knapsack shape worn on the back with an Indian rubber air-bag worn on the breast of the operator joined to the knapsack by air tubes. At the base of the knapsack was a cylinder containing oxygen sufficient to last four hours and a filter to extract the carbon dioxide exhaled by the wearer. Mr Fleuss, the inventor had agreed to lead the experiments and to form part of the exploring party to enter the Maudlin seam. The experiments began by isolating those wearing the apparatus in a sealed wooden erection in which a cauldron of burning sulphur had been placed filling the room with poisonous fumes. By the end of the experiment everyone was satisfied that the apparatus worked admirably in a noxious atmosphere. Fleuss also took the opportunity to test a patent lamp for use in case of emergency. That Friday evening all work at the pit ceased and all workmen brought their gear to bank. At 11.00am the following morning the horse keepers began to draw most of the pit ponies to bank in preparation for the removal of the stoppings. Some ponies were left underground in case they were needed by the exploration party. At the pithead a strong wooden barricade was erected around the shaft top to prevent the inquisitive and curious public or from relatives of the victims from interfering with the work of the officials at bank. Instructions were given that any bodies recovered were to be taken immediately to the "dead house" and after recognition, if it was possible to recognise them, they would be buried in the churchyard adjacent to the colliery.

Strict rules and instructions were also laid down for the exploring parties. No person was allowed in the pit without a special order. Each party of twelve men had an appointed leader who, with one companion, was the only person to test for gas with his lamp and the remainder of the party was to keep ten yards to the rear. Everyone else was told to be extremely careful with their lamps which were Davy's fitted in tin cases. The patent lamp of Mr Fleuss was also used. The lead members of the working parties wore the Fleuss breathing apparatus.

Mr Robert Barlow, Overman for the Maudlin District wearing the Fleuss Apparatus
Courtesy of Seaham Family History Group

Three of the breathing apparatus were to be used by the explorers and a further two were made ready at bank to use when the oxygen was exhausted in the first three. The work of taking down the brick walls was conducted in a cautious and systematic manner. First the stopping leading into the east way of the seam was taken down and the work of clearing away the debris was rapidly carried out. The first problem soon became apparent. Owing to the amount of gas in the workings the ordinary safety lamps had to be discarded and the men continued their work using Mr Fleuss' patent limelight lamp. The Fleuss apparatus allowed the men to work relatively uninhibited in the foul air and by the end of that night about one third of the distance to the fallen part of the seam had been travelled leaving about 1,000 yards to be explored. The principal work of some of the supporting teams was to clear away the debris and make the roofs and roadways as safe as possible behind the main exploration teams. Large quantities of gas were frequently detected but the teams did not come across any fires. The fires were expected to be found in the unexplored part of the seam. Accompanying Mr Fleuss was his foreman Mr J Halloran, a professional diver from London who had experience of fires in the coalfields of Wales. Those members of the working parties with the Fleuss limelight lamps penetrated 200 yards further into the seam than the others and two men proceeded further to the Maudlin engine plane to assess the conditions in that area. The exploration parties reached the great fall and it was expected that the first of the bodies would be found half-a-mile beyond that point and would not be reached for another seven to ten days.

The Medical Officer of Health of the Easington Board of Guardians (of which the New Seaham Parish was then a part) had issued new sanitary regulations in view of the particular circumstances at Seaham Colliery. Any recovered bodies were not to be taken to the houses of the bereaved relatives. The blinds of the cottages in the village were respectfully closed and there was a prevailing quietness around the colliery rows and streets. The rector, Reverend WA Scott and his curate visited the colliery to get the latest information on the events below ground so that he could be prepared to deliver services and burials as soon as the victims were found.

One week after the stoppings were removed the exploration parties were still engaged in clearing away the falls in the Maudlin seam. It was expected that within a few days they may reach the landing where the fires were raging before the seam was sealed off. The work progressed slowly when a fall was found at the stables at the end of the east way with debris about eight feet deep. Below the fall were a number of dead pit ponies and the stench was overwhelming. The stables gave access to other parts of the seam and attempts were made to explore those parts to see if there were any bodies in that part of the mine. By the 29th July the explorers had reached the part of the Maudlin seam that had clearly been engulfed by a large fire. The way ahead was strewn with fallen wreckage but the work continued in spaces sometimes too narrow to move and with limited supplies of air. The teams had worked in relays without one hours break over the previous week. It was becoming evident beyond doubt that a ferocious and extensive fire had raged in the seam for a long time after the explosion took place. A quantity of burnt shale; large stones completely calcined and several pieces of burnt timber roofing was taken to bank and placed on view in the colliery office. Not far from the stables coal was found to be completely "coked". The actions of the mining engineers in promptly and firmly insisting that the seam be sealed off were completely vindicated. As the work progressed forwards the ventilation of the mine steadily improved and the explorers were beginning to think that they may have travelled beyond some of the bodies who could have been buried under the debris of the roof falls.

CHAPTER 31

The bodies of the missing men are found

Day 328 to 353,   31st July to 26th August 1881

After five weeks of exploration the place was reached where the mining engineers believed the bodies of the twenty-eight men would probably be found. The bodies of Thomas Cummings, shifter and Joseph Cowey, chock drawer were found on 31st July in the west pony way leading from the east landing but they were only recognisable by their clothes and the numbers on their lamps. It seemed that Thomas Cummings had been trying to find a way out but not having succeeded he was returning when the afterdamp came upon him and he died in the position he was found. He was an old man of seventy years and had worked at the colliery for the last thirty years. Joseph Cowey was sitting upon a piece of timber and both had been overcome by the effects of afterdamp. The watch belonging to Thomas Cummings was found close to his body. It had stopped at 1.23am almost twenty-three hours after the explosion occurred and it was very black and rusted. Both bodies were immediately sent to bank. The following day one more body was discovered, Joseph Pickles, a shifter who had died of the effects of afterdamp. The bodies were wrapped up and sent to the shaft bottom where they were put in coffins. At bank the coffins were taken straight to the dead house to await identification. The Coroner, Crofton Maynard was informed of the discovery of the bodies and he immediately opened an inquiry at the New Seaham Inn in the presence of Mr T Bell, HM Mines Inspector, Mr Corbett, viewer and Mr Barrett, colliery manager. After being sworn in the jury adjourned to the dead house at the pit to view the bodies. Formal evidence of identification was given of the bodies. The Coroner told the jury that at that stage of the inquiry there was nothing further to be done. Mr Corbett informed the Coroner that he expected the work in the Maudlin seam would be almost complete within one month. The order for the burial of the bodies was signed and the inquest was then adjourned until 31st August.

The exploration parties came upon the bodies of more men in an old disused way. The poor men seemed to have gone into the old disused way for safety but were overcome by afterdamp. The bodies were in a blackened mummified condition but were not offensive until later when they were exposed for some time in the air. James Ovington, chock drawer; Benjamin Ward, stoneman; Nathaniel Brown, hewer and John Spry, pillarman were brought to bank on 1st August. The body of eighteen-year-old William Crossman, putter was discovered in an old "wall" in a crouched position with his arms folded over his breast, clasping his lamp. His body was brought up on 2nd August. The work of the exploration party continued relentlessly in search of the remaining twenty bodies. Four days later the working parties reached the spot where it was believed that the bulk of the victims would be found. The body of John Grey, hewer was soon discovered near the coal face. On Tuesday 9th August the body of sixty-seven-year-old Joseph Theobald, shifter was found and the following day the bodies of William Roxby, hewer; Henry Turnbull, hewer and Alfred Turner, putter were discovered. One week later the body of sixteen-year-old Joseph Waller was found seated in a refuge hole. He had been overcome by afterdamp.

The workmen at the colliery had not worked since they were called out of the pit to allow the exploration parties to begin the taking down of the stoppings in the Maudlin seam. Notwithstanding the fact that the task of the explorers in recovering the bodies in the Maudlin had not yet been completed the men intimated to the colliery management that they were willing to re-commence work. That offer was accepted and by 20th August 1881 the No. 1 pit, No. 3 Main Coal and No. 3 Hutton districts were back to full working.

The explorers carried on the work of clearing debris, propping and repairing roofs and searching under every roof fall as they pushed forward into the final parts of the Maudlin seam. On 26th August they came across an area known as the far end of the north landing in the east district of the Maudlin seam. It was filled with a large pool of water and from the offensive smell they suspected that some of the bodies may be found under the water. There was so much water that it would have been impossible to drain the pool and so they brought pumps to the scene and began pumping the water out. Late on Friday night the water was sufficiently lowered to find the gruesome remains of nine bodies. Pumping continued throughout the night and the next morning and by Saturday afternoon a further four bodies were recovered. The bodies were later identified as: -

- Benjamin Redshaw, hewer
- Samuel Wilkinson, hewer
- John Wilkinson, hewer
- William Strawbridge, hewer
- Henry Elesbury, deputy
- John W Redshaw, shifter
- James Walker, packer
- James Johnson, packer
- Silas Scrafton, packer
- George Sharp, stoneman
- Thomas Wright, shifter
- John Copeman, stoneman
- Edward Johnson, shifter

The bodies were badly decomposed and had been in the water so long that they were much swollen and had a bloated appearance. They could only be identified by their clothing, shoes, watches and the buttons on their waistcoats.

CHAPTER 32

The final inquest

31st August 1881

The adjourned second inquest was reopened on a Wednesday morning in the Londonderry Literary Institute by Coroner Crofton Maynard. The first witness called was Mr Charles Rollo Barrett the Seaham Colliery Manager. He told the jury that he had been engaged every day in exploring the Maudlin seam for the previous nine weeks since 25th June. Almost all of the first groups of bodies discovered in the Maudlin had been found beyond falls of stone. Those bodies found on the 26th August were floating in an accumulation of water which had gathered in the seam. The bodies recovered were not burnt and all the indications agreed by the mining engineers was that the explosion had gone into the Maudlin seam but it had not originated there. All the bodies had been found except a boy called John Whitfield and all efforts were being made to find this last victim of the explosion. The indications of the explosion found in the Maudlin were doors, stoppings and sheaves blown inbye and he could not suggest any new theory of the seat or the cause of the explosion. He informed the jury of the signs of extreme heat and fierce fire that the exploration party discovered once the Maudlin seam was opened and explored. He was entirely satisfied that the decision of the committee of mining engineers to seal off that part of the pit was a necessary step to take. A number of witnesses agreed that if they had not taken this action and had pushed forward then it might have resulted in more loss of life from the exploration party. He stated that as colliery manager he was satisfied that none of the bodies walled up in the Maudlin seam could have been saved. Mr Norman Wilkinson, Treasurer to the DMA agreed with Mr Barrett's evidence and told the jury he was quite convinced it was necessary to close up the seam. He also thought that everything had been done by the owners under such a calamity to provide for the safety of the men and to recover the bodies.

There were no further witnesses called and the jury were informed by the Coroner that there was no other evidence to be considered. The jury at once returned the same verdict as pronounced at the first inquest. The Coroner accordingly announced that Thomas Cummings and Joseph Cowey and others were killed by an explosion on the 8th September 1880 but as to the seat or the origin of the explosion there was no new evidence to determine the cause. The inquest was then closed.

Seven weeks later at the Michaelmas Quarter Sessions at Durham the Court granted an honorarium to Coroner Crofton Maynard. He was awarded £15 per annum for five years in consideration of the arduous task imposed upon him in consequence of the Seaham Colliery disaster.

CHAPTER 33

The Durham Miners Association

Messrs Foreman, Patterson and Wilkinson, the president, financial secretary and treasurer respectively of the Durham Miners Association received a letter of thanks from the Marquess of Londonderry, the owner of Seaham Colliery. This letter commended the three agents for descending the mine with the exploration parties immediately after the terrible explosion and also for the excellent service they rendered on behalf of the men working with the colliery management.

The DMA was little more than ten years old when the terrible explosion happened on 8th September 1880. They had fought many battles for Durham pitmen in such a short period of time. Immediately preceding the formation of the DMA, a number of important events took place that forged the solidarity of pitmen throughout the County. The Mines Act came into force in July 1861 providing that no boy could go down the mine under twelve years of age unless he could produce a certificate that he could read and write; that coal should be weighed to ensure the men were paid fairly for their labour and that the workmen should be at liberty to appoint a checkweighman. A calamity followed in 1862 with the Hartley Colliery tragedy which turned the minds of men and Parliament to the necessity of sinking two shafts for every mine and the need for the financial provision for widows and children from a "permanent fund" when a miner was permanently injured or killed in the mine. Soon after, a strike at Brancepeth Colliery known as the "Rocking Strike" was caused by the checkweighman employed by the colliery owner. The owner's checkweighman was paid a commission for every tub he rejected as too light because they were not filled to the top when they came to bank. As a consequence, the hewer used to rock the tub to settle the coal in case the jolting on the road out might cause the coals to settle below the rim of the tub. Thereafter, the demands of the workmen were payment by weight not by numbers of full tubs.

A strike at Wearmouth Colliery in 1869 could claim to be the real starting point of the DMA. The strike arose out of the conditions contained in the "Bond" of that year. The "Bond" was an annual event usually held in March in which all of the workmen were called to the colliery office where the manager would read over all of the conditions of labour for the next twelve months. Usually, the first hewers to sign up to the bond at that meeting were paid a bounty to encourage others to step forward and commit their labour for another year. Chosen men would often be bribed with a sovereign to incite or induce the men to sign the bond. This was a legally binding contract the terms of which were often imposed through a court of law. If a pitman broke his bond and went to work for another colliery part way through the year the manager would haul him before the magistrate. The rate offered at the annual bond signing at Wearmouth Colliery was significantly less than the previous year prompting the men to strike. The breaking of the bond brought four test cases summonsed to appear at Sunderland Magistrates Court in June 1869. They were charged under the Masters and Servants Act and could have faced a fine of £20 or three-months imprisonment. The defence of the four test cases was conducted by Mr Roberts, the "Pitmen's Attorney-General". The defence argued that one of the defendants was a "marksman", that is, he made a mark because he could not read or write, and that the bond was never read over to him. Eventually, the men told Mr Roberts that they wanted to be rid of the iniquitous bond and they collectively undertook to leave their employment and colliery houses within nine days. The 300 men formed in procession each man carrying his lamp and a copy of the colliery rules and marched to the pit where they handed in their lamps and the rules to the colliery overman. The whole of the workmen soon joined in the strike and the solidifying of the men's resolve gave impetus to the cause of unionism throughout the county and so the roots of the Durham Miners Association were planted. Thomas Burt the future Member of Parliament for Morpeth delivered one of the more notable speeches to more than 1,000 miners assembled at Thornley in October, 1869 advocating the creation of a union. He urged that "there were many reasons why men should be united: wages, better conditions and safety at work"

One of the consequences of the creation of the DMA was the formation of a similar organisation representing the coal owners. The first meeting of the "Coal Owners Association" was held in February 1872 and their first task was to open communications with the DMA. The Secretary wrote to Mr Crawford of the DMA as follows: -

Mr Crawford, my dear Sir – I am directed to inform you that at a large meeting of the representatives of the household coal collieries, held here last Saturday, it was resolved that it is considered desirable that a meeting should be held between the coal owners and a deputation of the workmen at one o'clock on Saturday the 17th instant, at the Coal Trade Office, to discuss the various questions now in agitation by the workmen, with a view to their adjustment. Will you be so kind as to acknowledge the receipt of this letter and let me have the names of the deputation who will attend? I beg to remain, dear Sir, very respectfully yours, Theo Wood Bunning, Secretary.

At the outset the employers intimated that they were perfectly willing to abolish the bond and instead to replace it with a monthly or fortnightly agreement. The preference of the men was to have a fortnightly agreement and a schedule of meetings were agreed in the future to discuss this and many other issues that had been the source of industrial action in the past. A photograph of the first

N Wilkinson   W H Patterson   M Thompson   T Ramsey   W Crawford   G Jackson   J Foreman   T Mitchinson
(Treasurer)   Vice-President   (President and Secretary)
               W Askew

Deputation of the Durham Miner's Association to the Coal Trade Office, Newcastle deputation from the DMA to meet the Coal Owners Association is reproduced above. (Wilson, 1907)

The unity and camaraderie of the miners and the DMA is demonstrated at the annual Durham Miner's Gala. District meetings had been organised in various parts of the County for some years prior but the very first Gala was held at Wharton Park, Durham on 12th August 1871. A sum of £20 was offered in three prizes for a brass band contest and other money prizes were offered for various sport and athletic competitions. Although there was a charge for admission it was estimated that between 4,000 and 5,000 miners paid to attend the "big meeting". The Chairman Mr W Crawford invited speakers from outside the DMA such as Messrs A McDonald and W Brown from the Staffordshire miners and Mr John Normansell from the Yorkshire miners on to the platform. These speakers were men who had done great work in the Trade Union movement. The local speakers included Mr WH Patterson, Mr Hendry, Mr T Ramsey, and Mr N Wilkinson. The Chairman's first words were "This is the first great Gala Day of the Durham Miner's Mutual Confident Association, and I only pray that it will not be the last." His prayers were answered. Every year since then it has been held on the race-course ground at Durham City. In the following year 1872 the first part of the procession came into Durham City at 7.30am and from first to last there was good order amongst the visitors. There were 180 collieries and a large number of the 32,000 members of the DMA were present. Each colliery proudly held their own banner aloft as they marched to the racecourse ground to the rousing sound of their own brass band. Within a few years a Miner's Service at the magnificent Norman Cathedral at Durham was incorporated in the Gala ritual. The numbers attending grew steadily each year. In its halcyon days the number of visitors to the Gala exceeded 300,000. It quickly became the preeminent annual event in the labour movement's calendar. Many would congregate below the crowded balcony of the Royal County Hotel to hear speeches delivered by stalwarts of the labour movement before marching to the race-course ground to the rousing music from the numerous brass bands.

Anyone who visits the Gala can feel a tremendous sense of pride radiating from the miners and ex-miners; pride in their communities; pride in their culture and pride in their history. The tradition of the Durham Miner's Gala has survived the closing of the last pit in County Durham and the virtual extinction of the coal mining industry in Britain. One of the most poignant traditions still upheld at the Gala is the moment a number of chosen brass bands strike up "Gresford" the miners' Hymn of Remembrance.

The DMA play "Gresford" at Redhills,
the Headqurters of the DMA in December 2018

It was written by Robert Saint, a "Geordie" pitman from Hebburn Colliery who worked all his life down the pit. He was also a very gifted and talented musician. His work "Gresford" was dedicated to the 265 Welsh miners who lost their lives in the Gresford Colliery explosion on 22nd September 1934. It was adopted as the miner's hymn and it is a tradition that Gresford is played before each band leaves their pit village and it is the last hymn played in Durham Cathedral on Gala Day. To every Seaham Colliery miner attending the Gala it has a very special significance. On the anniversary of the pit disaster the colliery banner is draped in black. A hush parades through the crowds of ex-miners and visitors as the music of Gresford reaches the ears of everyone on the race-course ground. The lament stirs the soul of pitmen all around the racecourse ground as their thoughts dwell on the men maimed and killed over the years in seams like the Hutton or Maudlin. Almost every colliery has had a mining disaster but the miners of Seaham Colliery have more cause than most to become solemn and listen with a heavy heart as "Gresford" the miner's hymn is played.

CHAPTER 34

One year of Hell

Day 365, 7th September 1881

The weary exploration teams continued their search beyond the drained-out pool at the far end of the north landing. They were searching for one more victim of the disaster; just one more body to find and lay to rest in the communal grave in Christchurch. Where was the body of John Whitfield? Had they passed his body in an area already explored and would they have to retrace their steps and search again under the stone and timber littered along the ways? After twelve long days a body was found under a fall in the north landing of the east way. It was the seventeen-year-old driver John Whitfield. The date was the 7th September 1881. He had been entombed for one year exactly. John Whitfield's family and the pit community had been overwhelmed by every emotion that a human being could experience - hope, despair, anguish, torment, grief and pain. All of the loved ones left behind by the victims of that dreadful explosion on 8th September 1880; the 107 widows, 2 mothers, 2 guardians and 249 fatherless children and the whole of the pit community had gone through the worst experience any mortal could bear. Now all of the bodies had been recovered and laid to rest in the church graveyard across the road from the pit. The families in the colliery rows and streets could try and get back to normal. Every street in the pit village (See chart below) housed a widow or a child of one of the victims of the disaster. One hundred and thirty-two men and boys out of the one hundred and sixty-four who perished in the explosion lived in the colliery rows and streets surrounding the pit. Ten per cent of the male population of the pit village died in one day. The others who perished lived in

Seaham Harbour with some living at Murton and Sunderland. The Seaham Colliery pit rows and houses were a constant reminder of that dreadful tragedy until they were finally demolished over 70 years later.

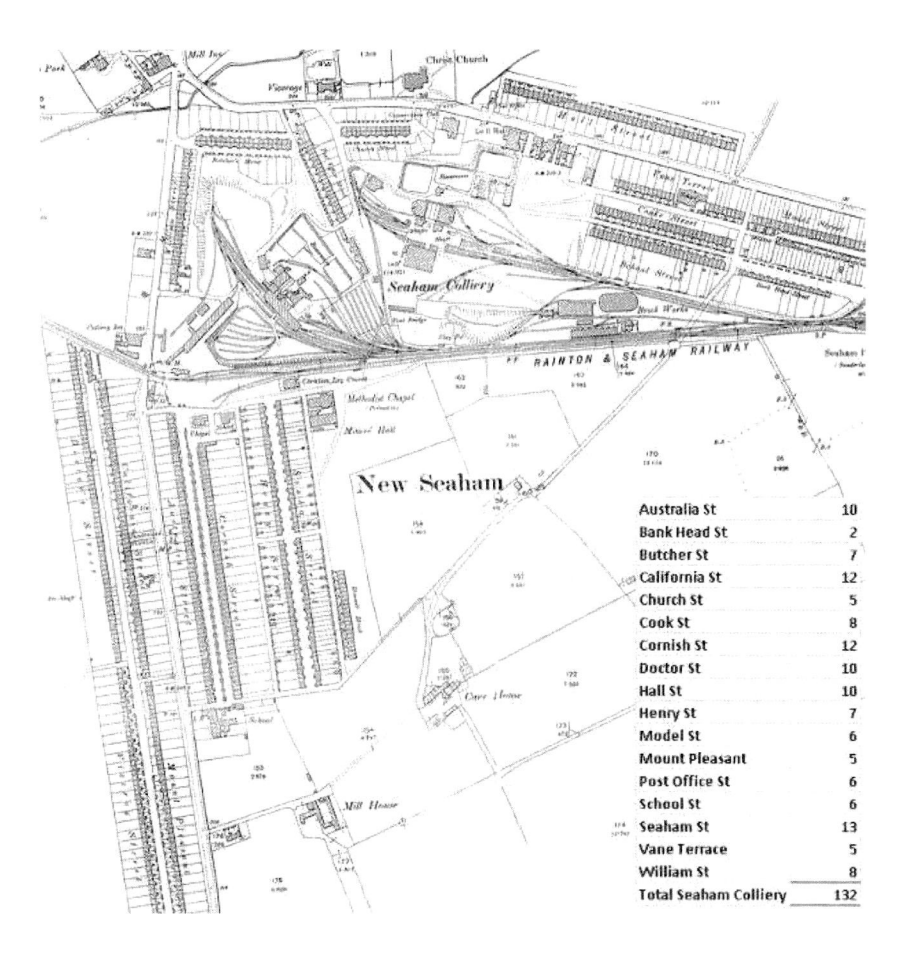

Australia St	10
Bank Head St	2
Butcher St	7
California St	12
Church St	5
Cook St	8
Cornish St	12
Doctor St	10
Hall St	10
Henry St	7
Model St	6
Mount Pleasant	5
Post Office St	6
School St	6
Seaham St	13
Vane Terrace	5
William St	8
Total Seaham Colliery	**132**

(Every pit row and street housed the family of one of the victims)
Chart displaying the number of victims by row or street

The discovery of John Whitfield's body and his burial service at Christchurch on 8th September 1881 supplied the closing incidents in a dark chapter in the history of Seaham. Everyone in the village remembered the shock and the terrible anxiety of pit families during the first two days after the disaster. The thoughts of wives and children alternating between hope and fear for the lives of the poor colliers caught in the fiery blast that rushed through the roadways of the mine. No-one could forget the anxious despairing faces that crowded around the pithead waiting for news or the selfless bravery of the men who ventured down the shaft to explore through the chaos and bring up the injured and the dead. The dangers of re-opening the Maudlin seam and the perseverance of the exploration parties until the last body was found deserve the highest praise for all who took part in it. It was without doubt a miracle in itself that the brave work carried out by the Seaham Colliery miners in recovering the bodies of their workmates was accomplished without any further sacrifice of life. It had truly been **"ONE YEAR OF HELL"**

CHAPTER 35

In Memoriam

Almost three years after the explosion at Seaham Colliery support was growing for the erection of some form of permanent monument to the men who died in the disaster. The officials and workmen at the colliery and families in the pit village needed a focal point to perpetuate the memory of the victims of the sad calamity. A committee was formed to inaugurate a fund for the purpose of buying and erecting a monument in Christchurch cemetery. Mr VW Corbett, Chief Agent at Seaham Colliery was appointed the Chairman. Fundraising began and a subscription list was circulated throughout the town headed by the Marquess of Londonderry who donated £10. Other subscriptions came flooding in to the committee from residents of the neighbourhood and adjacent towns.

Tenders and designs for a memorial were invited from stone masons and monumental carvers and Messrs C Ryder & Son of Bishop Auckland submitted a beautiful and artistic design and were the successful bidders for the work. Their design was a monument eighteen feet high built of freestone, with granite columns running up the corners and surmounted by a large cross. The names of the men and boys who lost their lives were carved into four slate slabs surrounding the monument and the inscription on the base read *"Erected by the workmen of Seaham and Rainton Collieries and other friends in memory of the 164 above-named men and boys who lost their lives in an explosion at Seaham Colliery on the 8th September 1880."* Surmounting the pedestal on each side were scripture texts that reminded the reader of the inevitability of death. They read: -

"Blessed are the dead which die in the Lord"

"He will swallow up death in victory"

"There's but a step between me and death"

"What man liveth and shall not see death"

Everyone who saw the statue expressed their great admiration for the work done by Messrs Ryder & Son. At the unveiling ceremony on 2nd June 1884 Canon Scott, Vicar of Christchurch gave a short service in the crowded church followed by an address by the Venerable Archdeacon Watkins of Durham. The choir and procession then moved to the platform near the monument where stood a number of invited clergy and guests including the Marquess of Londonderry, Viscount Castlereagh, Reverends' A Bethune, J Colling, J Wallis, J Copley and J James, Mr VW Corbett, Mr J Barnett, Mr Warham, Mr SJ Ditchfield and many others.

The Marquess of Londonderry gave a very moving and eloquent speech to the large crowds reported to have been in excess of 60,000 people listening attentively around the church and thronging the roads approaching the church. He said: -

"We have now met to perform the last sad offer of respect to our departed friends. The remembrance of the sad event that took place at Seaham Colliery three years ago is still very vividly in my mind, and I am sure in the mind of you all. Although, sad as it was it has been a source of very great satisfaction to me that I was present on the occasion, and endeavoured, as far as I could to render advice and assistance. You will also, I am sure, remember the very gracious response, the great sympathy displayed by Her Most Gracious Majesty, the Queen, who addressed us at the time in words of such deep feeling; and I can only say, on the part of Lady Londonderry, that she would have been only too happy to have been present today, and I am sure you will believe me when I say

Memorial to the victims of the 1880 Seaham Colliery Explosion

Courtesy of Alan Charlton

she felt the deepest sympathy on that occasion. I also remember I had the satisfaction of my son, Lord Castlereagh, being present. He took a great interest in the calamity, and came a long distance to endeavour to give such assistance as he possibly could. Well, gentlemen, we are now called together to perform the last ceremony, the last tribute of respect; but nothing I am sure will ever make us forget that fearful, sad morning, with its crushing horror, that first taught us that we had lost so many true and good men. Before unveiling this monument, I must express my great admiration of your most excellent Miners' Permanent Fund. It does them great credit that they have a fund with such a purpose; and I am glad it is so well supported by the miners of this County and of Northumberland. I am anxious to bear my testimony to this great institution, for no one appreciated more than I did the great benefit it was to those deprived of their breadwinners on that sad occasion. I will now unveil this monument to the memory of our departed friends with every sentiment of respect, remembrances, and devotion."

The memorial was then unveiled and several of the lady's placed wreaths around the base of the monument.

The Reverend Canon Scott thanked the Marquess for his kindness for attending and unveiling the monument and for the devotion he showed to his men in those bleak days after the calamity. He also thanked Lady Londonderry for the sympathy she expressed when she accompanied him on visits to the widows in their pit houses in the days following their bereavement. He thanked the fundraising committee for the splendid work they had done and the workmen at Seaham Colliery who had contributed much of the cost of the monument.

The shrine has since been incorporated into a Miner's Memorial Garden. The public can sit and gaze at the memorials to the victims of the 1871 and 1880 explosion and reflect on the lives and deaths of the poor souls listed on those solemn edifices. Sometimes a death can be a good death. When someone has lived a long life and has had a good life there is still the pain of parting but we know that the person's life has run its course. Usually, we have the opportunity to prepare ourselves for their death and even say goodbye. Sometimes it is a bad death, a cruel and premature death, particularly when there is a tragedy. Often the ones left behind feel betrayed or cheated. But what about the miners in the Hutton and Maudlin seam who knew beyond doubt that they were facing a premature death on that fateful September day? One can only imagine the thoughts of those pitiful men and boys whose last view of the earth and the sky before they went down the pit was peace and tranquillity. Then they entered the cage and descended down the shaft into a dark, damp gassy and hostile environment. At the coalface or other workplace, they became dependent for survival on their workmates. They had to entrust their existence to the people in their shift who entered the pit with them or worked side by side with them. This "camaraderie" is almost unique to the mining community. Every pitman knew when they went inbye that they may not come back to bank again but they just didn't dwell on it. Then, just a few hours after the start of their shift the men in the Hutton and Maudlin seams felt that fiery blast rushing through the mine and they knew that their mortal life on this earth was fading away. Many of the 164 men and boys died on 8th September 1880 or within a few days afterwards. The details of the date and place that their bodies were discovered and the date they were finally interred in consecrated ground are attached as Appendix 2. One hundred and twenty-nine bodies were laid to rest at Christchurch; twenty-three at St John's church; Five at St Mary's church and seven men were laid to rest at other churches in parishes outside of the town. Now in my sixtieth decade I often sit quietly on the seat in the Miners Memorial Garden at Christchurch on a bright, crisp morning and reflect on the 164 names of the victims carved into the four marble tablets. I ask myself why many thousands of Durham miners worked in such terrible places. Those miners went down again and again to earn money to feed their families because there was very little work elsewhere and because it was their legacy. Their father worked at the pit and as they were part of the pit village and community they were expected to follow in his footsteps. I am so thankful that my father emphatically declared when I was just a young lad that he would never allow me to work down the pit.

There was an ash dirt track running parallel with the railway lines at Seaham Colliery. It was called the "Black Road" and at the beginning or end of a shift pitmen would often walk to and from their homes along the black path. A poem epitomising the closing of the collieries and "black roads" at every pit was written by a Durham man.

The Pit Road

It is quiet on the pit road down the hill
Where once the gaitered legs
Shaped bishops in the gloom,
The clump of bartered boots is faintly heard
In ghostly winds which blow there still.

There is small happiness on this grey earth
And accidents of joy are rare
To keep one here, Whistle if you can
In echoes of the tankey tones
Where once the wheels stood grinning
With a soulless mirth.

My father died among the aching years
And when the north wind combs,

The hairy dust, I hear again
The hobbled hardness of his foot
Full on the metal of the old pit road
Winding along my tears

Only the ash remains
Of older pits in elephantine capes
Only the ash remains
Once recognised and loved in human form
The dead face emptied of its life and worth
Only the ash remains on mother earth

And yet I hope that there is more than this
That there are energies which flicker on
When all I know has vanished in the rain.

William Dowding

Seaham Colliery is now silent. The pit village rows and streets were demolished over seventy years ago and the black road is used no more. The pit was combined underground with Vane Tempest Colliery in 1988 and then closed four years later. The buildings were demolished and the site was completely cleared within a short time after its closure. For many years the once busy, noisy and grimy industrial brown field site stood empty as though the ground was sterile but then a few years ago construction of a new building was begun on the very ground where the colliers once walked. Seaham High School was officially opened in July 2018.

Seaham High School opened in July 2018

The New Testament describes hell as a place of fire, a bottomless pit, darkness, destruction and everlasting torment. These descriptors of hell take us to a place beyond the limits of our language to a place far worse than we could imagine but a place that those unfortunate victims of the explosion knew only too well. The victims of the disaster in September 1880 left 107 widows and 249 children in the town. After five generations the number of descendants of those children is estimated to be in the region of 4,000 many of whom still live in Seaham. There is no doubt that some of the pupils walking through the corridors of the High School are walking in the footsteps of their pitmen ancestors unaware of the "ONE YEAR OF HELL" in the history of Seaham.

THE END

PIT TERMINOLOGY & GLOSSARY

J.B Priestley visited County Durham in 1934 and he observed that "the local miners have a curious lingo of their own, which they call 'pitmatic'. It is, you might say, a dialect within a dialect, for it is used only by the pitmen when they are talking among themselves. When the pitmen are exchanging stories of colliery life, usually very grim stories, they do it in 'pitmatic', which is Scandinavian in origin, far nearer to the Norse than the ordinary Durham dialect."

Surviving pitmatic words represent a rich seam in the history of English dialect. Similarly, the names of tools, places and activities used down the pit are unique to the industry and are not used in any other occupation. More than that, in some instances a tool, occupation or activity may be called one thing at a particular pit and something entirely different at another pit. The young generation of today will never hear these words in day-to-day conversation from their parents as there are no longer any pitmen working in County Durham. Perhaps they will hear the odd word or two from their grandparents but these words and phrases are destined to die with that generation.

Afterdamp

A mixture of lethal gases after an explosion, colourless and deadly

Agent

The owner of the pit delegates his authority to the Agent who may have a number of separate collieries entrusted to him

Back-overman

One who supervises the management of the pit during the back shift i.e., from the time the overman leaves

Bank

The top of the pit

Banksman

Person in charge of the loading and unloading of the cage at the surface

Blackdamp

Term generally applied to carbon dioxide also known as stythe

Blower

A sudden discharge of inflammable gas from some chasm or fissure in the coal or stone

Bond

The annual agreement specifying the conditions upon which the colliers are hired

Brattice

Originally wooden partitions in a shaft whereby fresh air went down one part and foul or return air returned up another

Check weigher

The Coal Mines Regulation Act 1877 gave the miners the freedom to appoint checkweighman. They were paid by the men and were usually literate and union men.

Clanny lamp

An oil safety lamp invented by Dr Clanny

Collier

An experience miner

Colliery overman

A senior overman in charge of a shift reporting direct to an undermanager

Damp

Name given to various dangerous gases and derived from the German word "damph" meaning gas

Davy lamp

An oil safety lamp invented by Sir Humphrey Davy in 1816. It had a mesh gauze surrounding the flame that allowed methane gas to enter the lamp giving a warning sign to the carrier but prevented the flame from escaping back into the gaseous atmosphere. It gave off a very poor light.

Deputy

An official responsible to the overman for the management of a district of the pit. Their shift usually starts one or two hours before the hewers.

Door trapper

Usually, a boy opening and closing ventilation doors to allow tubs to pass through

Downcast shaft

Shaft which carries fresh air from the surface down to the mine workings

Engine wright

A person responsible for all mechanical issues on the surface and underground. The term later became mechanical engineer.

Firedamp

Methane gas, CH4, lighter than air, burns with a blue flame. It has an explosive range from 5% to 15% but it cannot burn above 15% because there would be insufficient oxygen in the air.

Furnace

The furnace was usually at the shaft bottom to create airflow. Eventually furnaces were prohibited and replaced by mechanical forms of ventilation.

Hewer

Coal face worker who cuts and loosens the coal with a pick. He was usually allocated to either the first shift or the back shift.

Inbye

The direction along a roadway towards the face or workings thus going away from the shaft bottom to work

Intake

Fresh air roadway leading to a coal face

Kibble

A large bucket on the end of a winding rope usually used in shaft sinking

Landing

A platform or place at the top or bottom of a shaft

Loose

Time when the men finished the end of their shift

Lodge

The local union branch

Marrow or "marra"

A partner or mate

Master-shifter

The person in charge of the shifters

Misfire

The complete or partial failure of a blasting charge

Onsetter

A person in charge of loading and unloading the cage at the bottom of the shaft. He signals to the banksmen and winding engineman when the cage is loaded with men or tubs and is ready to move.

Outbye

The direction along a roadway away from the working face and towards the shaft bottom.

Overman

The third rank of officers in the mine and in charge of a section of the mine.

He receives reports from the deputies and is responsible for the general management of the district in his charge reporting to the under-manager.

Putter

A boy employed to drag or pull loaded coal tubs

Rap

A bell signal to a winding engineman e.g. three raps to the winder meant men were ready to travel in the cage up the shaft

Return

Roadway along which the air travels from the face and out of the mine

Screens

At the pithead where coal is sorted from the dirt and stone before washing

Shaftsman

A person employed to work on the shaft. He could be a blacksmith, joiner, engineer or a bricklayer. This was intricate and dangerous work.

Shifter

A man who repairs the horse-ways and other passages in the mine and keeps them free from obstructions.

Shotfirer

A person in charge of explosives and who detonates shots of explosive.

Skeats

Wooden guides within which the cage slides up and down the shaft

Staple pit

A shaft underground down the pit connecting one level to another

Staple shaft

An underground shaft connecting two or more levels of the mine but not reaching the surface.

Stonedusting

The methodical spreading of a fine dust of crushed limestone otherwise known as calcium carbonate. The ignition of naturally occurring methane gas mixed with air and coal dust can cause a devastating explosion in a mine. To reduce the risk of this happening stonedusting was introduced. The concentration of suspended limestone dust particles in the path of the flame of an explosion would smother and arrest the progress of the blast. Stonedusting became compulsory on the 1st January 1921.

Stythe

Also known as blackdamp

Sump

The bottom of a shaft that is used as a collection point for drainage water

Trapper

Young boys employed to sit in the dark opening and closing the trap doors used to ventilate the pit

Undermanager

The person in charge of the underground mining operations reporting directly to the manager

Upcast shaft

Shaft which the air returns to the surface after ventilating the mine workings

Viewer

Person employed by the owner to oversee several mines

Wasteman

Generally older men employed in building pillars for the support of the roof in the waste and in keeping the airways open and in good order. The shifter is his assistant.

Winder

The winding engine that raises or lowers the cage in a shaft

APPENDIX 1a

NORTHUMBERLAND AND DURHAM MINERS PERMANENT RELIEF FUND
2nd October 1880

No	Name of Men and Boys Lost	Age	Name of Widow	Age	Number of Children	Legacies £	Fortnightly Allowance £ s d
1	Alexander, Thomas	36	Elizabeth Ann	36	4	5	1 6 0
2	Barress, William	28	Lucy	25	2	5	18 0
3	Batey, John	33	Jane Ann	32		5	10 0
4	Bell, William	44	Elizabeth	40	5	5	1 10 0
5	Berry, William	26	Mary Ann	25		5	10 0
6	Birkbeck, Joseph	61	Ann	64		5	10 0
7	Best, James	51	Jane	50	2	5	18 0
8	Bowden, Joseph	22	Single			23	
9	Breeze, William	33	Ann	34	5	5	1 10 0
10	Brown, James	54	Alice	56	1	5	14 0
11	Brown, Edward	21	Single			23	
12	Brown, George	62	Widower			23	
13	Brown, Nathaniel	20	Single			23	
14	Brown, George	26	Single			23	
15	Carroll, Patrick	41	Widower			23	
16	Cassidy, Thomas	22	Susannah	24	1	5	14 0
17	Chapman, Joseph	35	Elizabeth	35	4	5	1 6 0
18	Charlton, Matthew	29	Ann	28	3	5	1 2 0
19	Clark, James, Sen	47	Sarah Ann	50	3	5	1 2 0
20	Clark, James, Jnr	20	Single			23	
21	Clark, Robert	71	Isabella	72		5	10 0
22	Clark, Joseph	23	Margaret	21		5	10 0
23	Cole, Richard	44	Elizabeth	42	2	5	18 0
24	Copeman, John	32	Elizabeth	32		5	10 0
25	Cook, Joseph	32	Maria	24	3	5	1 2 0

APPENDIX 1b

No	Name of Men and Boys Lost	Age	Name of Widow	Age	Number of Children	Legacies £	Fortnightly Allowance £ s d
26	Cowey, Joseph	40	Mary Ann	46	5	5	1 10 0
27	Crossman, William	18	Single			23	
28	Cummings, Thomas	71	Rebecca	68		5	10 0
29	Dawson, Walter	49	Jane	46	2	5	18 0
30	Dawson, Charles	37	Jane Ann	32	6	5	1 14 0
31	Dawson, Robson	34	Ellen	29	3	5	1 2 0
32	Defty, Richard	28	Mary	21	1	5	14 0
33	Dinning, John	51	Wdwer/Guardn			5	7 0
34	Diston, George	55	Widower			23	
35	Ditchburn, Isaac	39	Sarah	39	7	5	1 18 0
36	Dixon, Lees Ball	26	Single			23	
37	Dixon, George	47	Hannah	48	2	5	18 0
38	Dotchin, James	62	Charlotte	64		5	10 0
39	Driver, Richard	56	Elizabeth	55	2	5	18 0
40	Dunn, Robert	24	Margaret	20	1	5	14 0
41	Elesbury, Henry	70	Mary	31	2	5	18 0
42	Fife, William	44	Sarah	43	4	5	1 6 0
43	Fletcher, Jacob	50	Jane	42	5	5	1 10 0
44	Forster, Thomas	17	Single			23	
45	Forster, Thomas	64	harriet	54	1	5	14 0
46	George, Richard	31	Wdwer/Guardn		5	5	1 15 0
47	Gibbons, Dominic	46	Widower			23	
48	Greenwell, Robert Thomas	29	Hannah	26	1	5	14 0
49	Greenbanks, Anthony	27	Esther	41	1	5	14 0
50	Grey, John	51	Elizabeth	60		5	10 0
51	Grounds, John	19	Single			23	
52	Grounds, Thomas	27	Hannah	28	4	5	1 6 0
53	Hall, William	60	Elizabeth	55		5	10 0
54	Hancock, William	19	Single			23	

207

APPENDIX 1c

No	Name of Men and Boys Lost	Age	Name of Widow	Age	Number of Children	Legacies £	Fortnightly Allowance £ s d
55	Haswell, Robert	19	Single			23	
56	Hayes, Thomas, Sen	46	Ann	46	2	5	18 0
57	Hayes, Thomas, Jnr	23	Eliza	21	1	5	14 0
58	Hedley, James	17	Half-member			12	
59	Henderson, Michael, Sen	57	Mary	51	1	5	14 0
60	Henderson, Roger	25	Single			23	
61	Henderson, Michael, Jnr	22	Single			23	
62	Henderson, William	19	Single			23	
63	Higginbottom, James	62	Ann	61		5	10 0
64	Hindson, Thomas	40	Jane	40	2	5	18 0
65	Hopper, George	51	Single			23	
66	Horan, Charles	26	Single			23	
67	Hood, William	28	Mary Ann	23	3	5	1 2 0
68	Hutchinson, Thomas	60	Dorothy	63	2	5	18 0
69	Jackson, John	61	Elizabeth	63		5	10 0
70	Johnson, Robert	34	Thomosin	34	4	5	1 6 0
71	Johnson, Edward	39	Barbara	42	3	5	1 2 0
72	Johnson, James	22	Single			23	
73	Keenan, Thomas	37	Mary Ann	38	2	5	18 0
74	Keenan, Michael	50	Mary Ann	50		5	10 0
75	Kent, James	16	Half-member			12	
76	Kirk, John	67	Single			23	
77	Knox, John	17	Half-member			12	
78	Knox, David	14	Half-member			12	
79	Lamb, George, F	36	Ann	35	5	5	1 10 0
80	Lawson, Robson	14	Half-member			12	
81	Lock, John	50	Elizabeth	49	1	5	14 0
82	Lonsdale, Joseph	67	Ellen	61		5	10 0
83	Lonsdale, Joseph	48	M	48	2	5	18 0

APPENDIX 1d

No	Name of Men and Boys Lost	Age	Name of Widow	Age	Number of Children	Legacies £	Fortnightly Allowance £ s d
84	Lonsdale, John	27	Mary Ann	27	2	5	18 0
85	Lowdey, Thomas	48	Mary	48		5	10 0
86	Mason, John	21	Single			23	
87	McGuiness, John	31	Single			23	
88	McLoughlin, James William	54	Isabella	59		5	10 0
89	Miller, John	25	Mary	23	2	5	18 0
90	Moore, William	30	Margaret	30	3	5	1 2 0
91	Morris, William	54	Elizabeth	62		5	10 0
92	Murray, Walter	42	Elizabeth	41	5	5	1 10 0
93	Neasham, John	42	Mary Ann	42		5	10 0
94	Norris, George Henry	23	Single			23	
95	Ovington, James	49	Widower			23	
96	Owens, John	16	Half-member			12	
97	Owens, Michael	14	Half-member			12	
98	Page, George	55	Elizabeth	53		5	10 0
99	Patterson, John Thomas	31	Mary Jane	28	4	5	1 6 0
100	Phillips, Mark	22	Mary Ellen	22	1	5	14 0
101	Pickles, Joseph	51	Mary Ann	42	2	5	18 0
102	Pinkard, Edward	18	Half-member			12	
103	Potter, John	43	Margaret	37	4	5	1 6 0
104	Potter, Robert	47	Dorothy	46	2	5	18 0
105	Potts, William	41	Ellen	45	4	5	1 6 0
106	Ramshaw, Anthony	65	Ann	65		5	10 0
107	Rollins, Robert	39	Margaret Ann	40	6	5	1 14 0
108	Rollins, Joseph	49	Margaret	43	2	5	18 0
109	Redshaw, Wm John	22	Single			23	
110	Redshaw, Benjamin	25	Mary	25	2	5	18 0
111	Richardson, William	53	Hannah	53		5	10 0
112	Riley, John	70	Single			23	

APPENDIX 1e

No	Name of Men and Boys Lost	Age	Name of Widow	Age	Number of Children	Legacies £	Fortnightly Allowance £ s d
113	Roberts, Thomas	45	Ann	48	2	5	18 0
114	Roper, George	50	Isabella	44	4	5	1 6 0
115	Roper, John George	25	Mary Jane	19		5	10 0
116	Roxby, William	24	Elizabeth	28	4	5	1 6 0
117	Sanderson, Alexander	34	Mary Ann	34	6	5	1 14 0
118	Scarfe, Anthony	40	Ann	40	4	5	1 6 0
119	Scrafton, Silas	17	Single			23	
120	Sharp, George	39	Margaret	40	4	5	1 6 0
121	Shields, George	23	Single			23	
122	Shields, James	25	Mary Ann	23		5	10 0
123	Shields, Robert	52	Ann	48		5	14 0
124	Short, John	18	Single			23	
125	Sawey, William	30	Mary	24	3	5	1 2 0
126	Simpson, William	31	Margaret	30	2	5	18 0
127	Slavin, James	26	Cathrine	23	1	5	14 0
128	Smith, Michael	34	Margaret	33	3	5	1 2 0
129	Smith, Anthony	39	Mary	30	5	5	1 10 0
130	Smith, Christopher	36	Mary Ann	28	1	5	14 0
131	Smith, Thomas	50	Mary Jane	59		5	10 0
132	Smith, Luke	26	Frances	22	2	5	18 0
133	Sutherland, John	40	Elizabeth	33	9	5	2 6 0
134	Spanton, William	39	Ann	38	5	5	1 10 0
135	Spry, John	49	Sarah	50	2	5	18 0
136	Straughan, Joseph	21	Mother/ 2 children		2	5	18 0
137	Straughan, Robt Clark	17	Single			23	
138	Strawbridge, William	48	Margaret	46	3	5	1 2 0
139	Taylor, Wm Henry	24	Single			23	
140	Theobald, Joseph	67	Sarah	61		5	10 0
141	Turnbull, Henry (Bleasdale)	23	Jane	23	3	5	1 2 0

210

APPENDIX 1f

No	Name of Men and Boys Lost	Age	Name of Widow	Age	Number of Children	Legacies £	Fortnightly Allowance £ s d
142	Turner, Alfred Jas	18	Single			23	
143	Venner, Samuel	52	Joanah	45		5	14 0
144	Venner, William	24	Single			23	
145	Vickers, John	52	Hannah	52	1	5	14 0
146	Waller, Joseph	16	Half-member			12	
147	Walker, James	42	Jane	40	4	5	1 6 0
148	Ward, Benjamin	35	Sarah	32	4	5	1 6 0
149	Watson, John	38	Single			23	
150	Watson, Frank	61	Jane	51	1	5	14 0
151	Weir, John	47	Jane	42	6	5	1 14 0
152	Wharton, Robert	35	Margaret	30	2	5	18 0
153	Wilkinson, William	40	Single			23	
154	Wilkinson, Samuel	26	Elizabeth	26	3	5	1 2 0
155	Wilkinson, John	20	Mother	55		5	10 0
156	Wilkinson, William	20	Single			23	
157	Williams, Thomas Henry	14	Half-member			12	
158	Williams, George David	18	Single			23	
159	Whitfield, John	17	Half-member			12	
160	Wright, Thomas	26	Mary Ann	27	3	5	1 2 0
					246	1579	104 2 0

The following were not members of the Fund (Paid on behalf of the Seaham Relief Committee)

1	Gibson, Thomas	37	Mary	37	3	5	1 2 0
2	Graham, Robert	26	Single			4	
3	Hunter, John	25	Single			5	
4	Ramshaw, Henry	33	Single			4	
					3	18	1 2 0

APPENDIX 2a

Details of bodies recovered from the mine and interred

Name of Men and Boys lost	Age	Occupation	Date body found	Location body found	Cause of death if known	Christchurch	St John's	St Mary's	Murton
Brown, James	54	Stoneman	09/09/1880	No. 3 Hutton	Burnt	11/09/1880			
Simpson, William	31	Stoneman	09/09/1880	Maudlin curve	Burnt	11/09/1880			
Alexander, Thomas	36	Deputy	09/09/1880	No. 1 Hutton	Mutilated	12/09/1880			
Breeze, William	33	Horsekeeper	11/09/1880	Maudlin stables	Not known	12/09/1880			
Dawson, Walter	49	Furnaceman	09/09/1880	High pit furnace	Burnt	12/09/1880			
Dixon, Lees Ball	26	Hewer	09/09/1880	No. 1 Hutton	Afterdamp	12/09/1880			
Dixon, George	47	Shifter	09/09/1880	Bk staple Hutt No 3	Afterdamp		12/09/1880		
Dotchin, James	62	Shifter	09/09/1880	Sth shaft of No.1 pit	Mutilated	12/09/1880			
Forster, Thomas	64	Shifter	09/09/1880	No. 1 Hutton	Not known	12/09/1880			
George, Richard	31	Hewer	09/09/1880	No. 1 Hutton	Not known	12/09/1880			
Gibson, Thomas	37	Fireman	09/09/1880	No. 3 furnace	Mutilated		12/09/1880		
Hunter, John	25	Furnaceman	09/09/1880	No. 1 Hutton	Burnt		12/09/1880		
Kent, James	16	Putter	09/09/1880	No. 1 Hutton	Not known		12/09/1880		
Lawson, Robson	14	Driver	09/09/1880	Bk staple Hutt No 3	Afterdamp	N/A			
Lowdey, Thomas	48	Deputy	09/09/1880	No. 3 pit staple	Mutilated	12/09/1880			
Mason, John	21	Hewer	09/09/1880	Bk staple Hutt No 3	Mutilated	12/09/1880			
McGuiness, John	31	Hewer	09/09/1880	No. 3 Hutton	Mutilated	12/09/1880			
Neasham, John	42	Horsekeeper	10/09/1880	Stable No. 3 Hutton	Not known	12/09/1880			
Patterson, John T	31	Shifter	09/09/1880	No. 1 Hutton	Mutilated	12/09/1880			
Potter, Robert	47	Deputy	09/09/1880	No. 3 pit staple	Burnt	12/09/1880			
Ramshaw, Anthony	65	Shifter	11/09/1880	Shaft to Polka way	Mutilated	12/09/1880			
Rollins, Joseph	49	Shifter	09/09/1880	High pit staple	Mutilated	12/09/1880			
Rollins, Robert	39	Stoneman	09/09/1880	High pit staple	Not known	12/09/1880			
Smith, Anthony	39	Master Shifter	09/09/1880	High pit staple	Not known	12/09/1880			
Spanton, William	39	Horsekeeper	09/09/1880	No. 1 Hutton	Mutilated	12/09/1880			
Straughan, Joseph	21	Hewer	09/09/1880	No. 3 Hutton	Not known	12/09/1880			

APPENDIX 2b

Details of bodies recovered from the mine and interred

Name of Men and Boys lost	Age	Occupation	Date body found	Location body found	Cause of death if known	Christchurch	St John's	St Mary's	Murton
Venner, Samuel	52	Stoneman	09/09/1880	No. 3 pit staple	Mutilated	12/09/1880			
Venner, William	24	Stoneman	09/09/1880	No. 3 pit staple	Not known	12/09/1880			
Weir, John	47	Deputy	09/09/1880	No. 1 Hutton	Mutilated	12/09/1880			
Wilkinson, William	20	Putter	09/09/1880	High pit staple	Mutilated	12/09/1880			
Williams, Thomas H	14	Driver	09/09/1880	High pit staple	Not known	12/09/1880			
Chapman, Joseph	35	Deputy	09/09/1880	High pit staple	Mutilated	12/09/1880			
Hindson, Thomas	40	Stoneman	13/09/1880	Return in Maudlin seam	Mutilated	13/09/1880			
Charlton, Matthew	29	Hewer	17/09/1880	Far-off way Hutt No 1	Afterdamp		14/09/1880		
Forster, Thomas	17	Putter	17/09/1880	Far-off way Hutt No 1	Afterdamp	18/09/1880	18/09/1880		
Gibbons, Dominic	46	Shifter	17/09/1880	Far-off way Hutt No 1	Afterdamp	18/09/1880			
Graham, Robert	26	Hewer	17/09/1880	Far-off way Hutt No 1	Afterdamp	18/09/1880			
Greenwell, Rbt Thomas	29	Deputy	17/09/1880	Far-off way Hutt No 1	Afterdamp	18/09/1880			
Henderson, Roger	25	Shifter	17/09/1880	Far-off way Hutt No 1	Afterdamp	18/09/1880			
Henderson, M, Jnr	22	Shifter	17/09/1880	Far-off way Hutt No 1	Afterdamp	18/09/1880			
Henderson, William	19	Putter	17/09/1880	Far-off way Hutt No 1	Afterdamp	18/09/1880			
Jackson, John	61	Shifter	17/09/1880	Far-off way Hutt No 1	Afterdamp	18/09/1880	18/09/1880		
Owens, Michael	14	Driver	17/09/1880	Far-off way Hutt No 1	Afterdamp	18/09/1880			
Shields, George	23	Hewer	17/09/1880	Far-off way Hutt No 1	Afterdamp	18/09/1880			
Short, John	18	Shifter	17/09/1880	Far-off way Hutt No 1	Afterdamp	18/09/1880			
Slavin, James	26	Hewer	17/09/1880	Far-off way Hutt No 1	Afterdamp		18/09/1880		
Horan, Charles	26	Hewer	17/09/1880	Far-off way Hutt No 1	Afterdamp	19/09/1880			
Richardson, William	53	Hewer	17/09/1880	Far-off way Hutt No 1	Afterdamp			19/09/1880	
Watson, Frank	61	Shifter	17/09/1880	Far-off way Hutt No 1	Afterdamp		19/09/1880		
Williams, George D	18	Putter	17/09/1880	Far-off way Hutt No 1	Afterdamp		19/09/1880		
Barress, William	28	Hewer	20/09/1880	3rd East Way	Afterdamp		20/09/1880		
Clark, James, Sen	47	Hewer	20/09/1880	3rd East Way	Afterdamp	20/09/1880			
Clark, James, Jnr	20	Hewer	20/09/1880	3rd East Way	Afterdamp	20/09/1880			

APPENDIX 2c

Details of bodies recovered from the mine and interred

Name of Men and Boys lost	Age	Occupation	Date body found	Location body found	Cause of death if known	Christchurch	St John's	St Mary's	Murton
Fletcher, Jacob	50	Shifter	19/09/1880	No. 1 Hutton	Afterdamp	20/09/1880			
Haswell, Robert	19	Putter	20/09/1880	3rd East Way	Afterdamp		20/09/1880		
McLoughlin, James W	54	Chock Drawer	19/09/1880	No. 1 Hutton	Afterdamp	20/09/1880			
Bell, William	44	Shifter	20/09/1880	3rd East Way	Afterdamp	21/09/1880			
Birkbeck, Joseph	61	Shifter	20/09/1880	3rd East Way	Afterdamp	21/09/1880			
Clark, Robert	71	Shifter	20/09/1880	3rd East Way	Afterdamp	21/09/1880			
Dawson, Charles	37	Master Shifter	20/09/1880	3rd East Way	Afterdamp	21/09/1880			
Driver, Richard	56	Packer	19/09/1880	No. 1 Hutton	Afterdamp	21/09/1880			
Fife, William	44	Stoneman	20/09/1880	3rd East Way	Afterdamp	21/09/1880			
Greenbanks, Anthony	27	Stoneman	20/09/1880	3rd East Way	Afterdamp	21/09/1880			
Hayes, Thomas, Sen	46	Stoneman	20/09/1880	3rd East Way	Afterdamp	21/09/1880			
Hayes, Thomas, Jnr	23	Stoneman	20/09/1880	3rd East Way	Afterdamp	21/09/1880			
Hedley, James	17	Putter	20/09/1880	3rd East Way	Afterdamp		21/09/1880		
Henderson, M, Sen	57	Shifter	20/09/1880	3rd East Way	Afterdamp	21/09/1880			
Hopper, George	51	Shifter	20/09/1880	3rd East Way	Afterdamp	21/09/1880			
Hood, William	28	Shifter	20/09/1880	3rd East Way	Afterdamp	21/09/1880			
Hutchinson, Thomas	60	Shifter	20/09/1880	3rd East Way	Afterdamp		21/09/1880		
Keenan, Michael	50	Shifter	20/09/1880	3rd East Way	Afterdamp	21/09/1880			
Kirk, John	67	Shifter	20/09/1880	3rd East Way	Afterdamp	21/09/1880			
Knox, John	17	Driver	20/09/1880	3rd East Way	Afterdamp			21/09/1880	
Knox, David	14	Driver	20/09/1880	3rd East Way	Afterdamp			21/09/1880	
Lamb, George, F	36	Stoneman	19/09/1880	3rd East Way	Mutilated	21/09/1880			
Lock, John	50	Stoneman	20/09/1880	3rd East Way	Afterdamp	21/09/1880			
Lonsdale, John	27	Hewer	20/09/1880	3rd East Way	Afterdamp	21/09/1880			
Norris, George H	23	Shifter	20/09/1880	3rd East Way	Afterdamp	21/09/1880			
Page, George	55	Stoneman	20/09/1880	3rd East Way	Afterdamp	21/09/1880			
Phillips, Mark	22	Stoneman	20/09/1880	3rd East Way	Afterdamp	21/09/1880			

APPENDIX 2d

Details of bodies recovered from the mine and interred

Name of Men and Boys lost	Age	Occupation	Date body found	Location body found	Cause of death if known	Christchurch	St John's	St Mary's	Murton
Pinkard, Ed (Burns)	18	Putter	20/09/1880	3rd East Way	Afterdamp		N/A		
Potts, William	41	Deputy	20/09/1880	3rd East Way	Afterdamp	21/09/1880			
Ramshaw, Henry	33	Shifter	19/09/1880	3rd East Way	Mutilated	21/09/1880			
Riley, John	70	Shifter	19/09/1880	No. 1 Hutton	Afterdamp	21/09/1880			
Roper, George	50	Hewer	20/09/1880	3rd East Way	Afterdamp	21/09/1880			
Sanderson, Alexander	34	Stoneman	20/09/1880	3rd East Way	Afterdamp	21/09/1880			
Scarfe, Anthony	40	Brakesman	20/09/1880	3rd East Way	Afterdamp	21/09/1880			
Shields, Robert	52	Shifter	20/09/1880	3rd East Way	Afterdamp	21/09/1880			
Smith, Thomas	50	Shifter	20/09/1880	3rd East Way	Afterdamp		21/09/1880		
Smith, Luke	26	Hewer	20/09/1880	3rd East Way	Afterdamp		21/09/1880		
Straughan, Robert C	17	Putter	20/09/1880	3rd East Way	Afterdamp	21/09/1880			
Hall, Edward	60	Chock Drawer	23/09/1880	Foot of Maudlin incline	Afterdamp	28/09/1880			
Morris, William	54	Chock Drawer	23/09/1880	Foot of Maudlin incline	Afterdamp	N/A			
Batey, John	33	Deputy	24/09/1880	The Maudlin incline	Afterdamp	29/09/1880			
Carroll, Patrick	41	Hewer	24/09/1880	The Maudlin incline	Afterdamp	N/A			
Cassidy, Thomas	22	Hewer	23/09/1880	Foot of Maudlin incline	Afterdamp		29/09/1880		
Clark, Joseph	23	Stoneman	24/09/1880	The Maudlin incline	Afterdamp	29/09/1880			
Dinning, John	51	Packer	24/09/1880	The Maudlin incline	Afterdamp	29/09/1880			
Diston, George	55	Shifter	24/09/1880	The Maudlin incline	Afterdamp	29/09/1880			
Grounds, John	19	Packer	24/09/1880	The Maudlin incline	Afterdamp	29/09/1880			
Grounds, Thomas	27	Packer	24/09/1880	The Maudlin incline	Afterdamp	29/09/1880			
Lonsdale, Joseph	48	Stoneman	23/09/1880	Foot of Maudlin incline	Afterdamp	29/09/1880			
Miller, John Thomas	25	Hewer	24/09/1880	Foot of Maudlin incline	Afterdamp	29/09/1880			
Murray, Walter	42	Master Shifter	23/09/1880	Foot of Maudlin incline	Afterdamp		29/09/1880		
Owens, John	16	Driver	24/09/1880	The Maudlin incline	Afterdamp		29/09/1880		
Sawey, William	30	Packer	23/09/1880	Foot of Maudlin incline	Afterdamp	29/09/1880			
Smith, Christopher	36	Shifter	24/09/1880	The Maudlin incline	Afterdamp		29/09/1880		

APPENDIX 2e

Details of bodies recovered from the mine and interred

Name of Men and Boys lost	Age	Occupation	Date body found	Location body found	Cause of death if known	Date Interred Christchurch	St John's	St Mary's	Murton
Taylor, Wm Henry	24	Packer	23/09/1880	Foot of Maudlin incline	Afterdamp	N/A			
Wilkinson, William	40	Shifter	23/09/1880	Foot of Maudlin incline	Afterdamp	29/09/1880			
Berry, William	26	Packer	24/09/1880	The Maudlin incline	Afterdamp	30/09/1880			
Best, James	51	Stoneman	23/09/1880	Foot of Maudlin incline	Afterdamp	30/09/1880			
Brown, Edward	21	Hewer	24/09/1880	The Maudlin incline	Afterdamp	30/09/1880			
Brown, George	62	Shifter	24/09/1880	The Maudlin incline	Afterdamp	30/09/1880			
Brown, George	26	Shifter	23/09/1880	Foot of Maudlin incline	Afterdamp	30/09/1880			
Cook, Joseph	32	Hewer	24/09/1880	The Maudlin incline	Afterdamp	30/09/1880			
Dawson, Robson	34	Deputy	24/09/1880	The Maudlin incline	Afterdamp	30/09/1880			
Defty, Richard	28	Packer	23/09/1880	Foot of Maudlin incline	Afterdamp	30/09/1880			
Ditchburn, Isaac	39	Packer	24/09/1880	The Maudlin incline	Afterdamp	30/09/1880			
Higginbottom, James	62	Shifter	23/09/1880	Foot of Maudlin incline	Afterdamp	30/09/1880			
Johnson, Robert	34	Packer	24/09/1880	The Maudlin incline	Afterdamp	30/09/1880			
Keenan, Thomas	37	Packer	24/09/1880	The Maudlin incline	Afterdamp	30/09/1880			
Lonsdale, Joseph	67	Shifter	23/09/1880	Foot of Maudlin incline	Afterdamp	30/09/1880			
Potter, John	43	Shifter	24/09/1880	The Maudlin incline	Afterdamp	30/09/1880			
Roper, John George	25	Hewer	23/09/1880	Foot of Maudlin incline	Afterdamp	30/09/1880			
Shields, James	25	Hewer	24/09/1880	The Maudlin incline	Afterdamp	30/09/1880			
Sutherland, John	40	Chock Drawer	23/09/1880	Foot of Maudlin incline	Afterdamp	30/09/1880			
Watson, John	38	Shifter	25/09/1880	Straight way of Maudlin	Afterdamp				30/09/1880
Wharton, Robert	35	Shifter	24/09/1880	The Maudlin incline	Afterdamp	N/A			
Bowden, Joseph	22	Hewer	25/09/1880	Straight way of Maudlin	Afterdamp		01/10/1880		
Cole, Richard	44	Packer	25/09/1880	Straight way of Maudlin	Afterdamp		01/10/1880		
Dunn, Robert	24	Stoneman	25/09/1880	Straight way of Maudlin	Afterdamp	01/10/1880			
Hancock, William	19	Putter	25/09/1880	Straight way of Maudlin	Afterdamp	01/10/1880			
Moore, William	30	Packer	25/09/1880	Straight way of Maudlin	Afterdamp	01/10/1880			
Roberts, Thomas	45	Stoneman	25/09/1880	Straight way of Maudlin	Afterdamp	01/10/1880			

APPENDIX 2f

Details of bodies recovered from the mine and interred

Name of Men and Boys lost	Age	Occupation	Date body found	Location body found	Cause of death if known	Christchurch	St. John's	St Mary's	Murton
Smith, Michael	34	Packer	25/09/1880	Straight way of Maudlin	Afterdamp	01/10/1880			
Vickers, John	52	Horsekeeper	01/10/1881	East way Maudlin seam	Mutilated	02/10/1880			
Cummings, Thomas	71	Shifter	31/07/1881	Pony way east landing	Afterdamp	01/08/1881			
Cowey, Joseph	40	Chock Drawer	31/07/1881	Pony way east landing	Afterdamp	02/08/1881			
Ovington, James	49	Chock Drawer	01/08/1881	Disused way in Maudlin	Afterdamp	02/08/1881			
Pickles, Joseph	51	Shifter	01/08/1881	Face of Maudlin seam	Afterdamp	02/08/1881			
Brown, Nathaniel	20	Hewer	01/08/1881	Disused way in Maudlin	Afterdamp	03/08/1881			
Crossman, William	18	Putter	02/08/1881	Disused way in Maudlin	Afterdamp	03/08/1881			
Spry, John	49	Packer	01/08/1881	Disused way in Maudlin	Afterdamp	03/08/1881			
Ward, Benjamin	35	Stoneman	01/08/1881	Disused way in Maudlin	Afterdamp	03/08/1881			
Grey, John	51	Hewer	05/08/1881	Face of Maudlin seam	Afterdamp	06/08/1881			
Theobald, Joseph	67	Shifter	09/08/1881	Face of Maudlin seam	Afterdamp	11/08/1881			
Roxby, William	24	Packer	10/08/1881	Face of Maudlin seam	Afterdamp	13/08/1881			
Turnbull, H (Bleasdale)	23	Stoneman	10/08/1881	Face of Maudlin seam	Afterdamp	13/08/1881			
Turner, Alfred Jas	18	Putter	10/08/1881	Face of Maudlin seam	Afterdamp	13/08/1881			
Waller, Joseph	16	Putter	17/08/1881	Face of Maudlin seam	Afterdamp	19/08/1881			
Elesbury, Henry	70	Deputy	26/08/1881	Nth landing east Maudlin	Afterdamp	28/08/1881			
Johnson, Edward	39	Shifter	26/08/1881	Nth landing east Maudlin	Afterdamp	28/08/1881			
Redshaw, Wm John	22	Shifter	26/08/1881	Nth landing east Maudlin	Afterdamp	28/08/1881			
Redshaw, Benjamin	25	Hewer	26/08/1881	Nth landing east Maudlin	Afterdamp	28/08/1881			
Scrafton, Silas	17	Packer	26/08/1881	Nth landing east Maudlin	Afterdamp		28/08/1881		
Sharp, George	39	Stoneman	26/08/1881	Nth landing east Maudlin	Afterdamp			28/08/1881	
Walker, James	42	Packer	26/08/1881	Nth landing east Maudlin	Afterdamp	28/08/1881			
Wilkinson, Samuel	26	Hewer	26/08/1881	Nth landing east Maudlin	Afterdamp	28/08/1881			
Wilkinson, John	20	Hewer	26/08/1881	Nth landing east Maudlin	Afterdamp	28/08/1881			
Copeman, John	32	Stoneman	26/08/1881	Nth landing east Maudlin	Afterdamp	29/08/1881			
Johnson, James	22	Packer	26/08/1881	Nth landing east Maudlin	Afterdamp	29/08/1881			

Details of bodies recovered from the mine and interred

Name of Men and Boys lost	Age	Occupation	Date body found	Location body found	Cause of death if known	Date Interred Christchurch	St John's	St Mary's	Murton
Strawbridge, Wm	48	Hewer	26/08/1881	Nth landing east Maudlin	Afterdamp		29/08/1881		
Wright, Thomas	26	Stoneman	26/08/1881	Nth landing east Maudlin	Afterdamp	29/08/1881			
Whitfield, John	17	Driver	07/09/1881	Nth landing east Maudlin	Afterdamp	08/09/1881			

APPENDIX 2g

BIBLIOGRAPHY AND REFERENCES

Bibliography:

Durham County Environmental Curriculum Group (1993) *Coal Mining in County Durham*

Mrs E Smith (1898) *Memoirs of a Highland Lady*

Lucinda J Fowler (1982) *Marriage, mining and mobility: four Durham parishes* published by Durham University

William Fordyce (1860) *A History of Coal, Coke, Coal Fields* published by Sampson Low & Son, London

T Fordyce (1867) *Local Records of Historical Register of Remarkable Events*

John J Atkinson (1864) *HM Inspector of Mines Report, Southern District*

John E McCutcheon (1955) *Troubled Seams* published by J Greenwood & Sons, Seaham

John Wilson (1907) *A History of the Durham Miner's Association* published by JH Veitch & Son

WN & JB Atkinson (1886) *Explosions in Coal Mines* published by Longmans, Green and Company

Francis Whellan (1894) *History, Topography and Directory of the County Palatine of Durham* 2nd Edition

References:

Richard Feynes (1873) *The Miners of Northumberland and Durham*

Volume 15 of Transactions of the North of England Institute of Mining Engineers

Trade Directories (various dates) *Pigots, Whites, Hagars, Slaters, Wards, Kellys, Whellans.*

Local Newspapers (various dates) *Seaham Weekly News, Seaham Observer, Sunderland Echo, Newcastle Courant, Newcastle Journal, Shields Gazette, Durham Chronicle, Hartlepool Mail, The Illustrated London News, The Advertiser, The Times, Reynolds Newspaper*

OTHER BOOKS BY THE AUTHOR

Please visit your favourite eBook retailer to discover other books by Fred Cooper:

THE HOLE-IN-THE-WALL:

This book is a social history study of a maritime town on the north-east coast of England during the 19th century. The mariners and the sailing ships entering harbour on each census day from 1831 to 1911 are analysed and the results present a clear picture of the lives of mariners in Victorian times.

A MASTER MARINER'S TALE: A Victorian Action Adventure and Spy Thriller Novel:

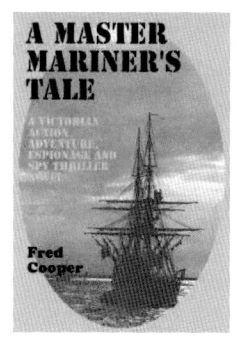

The town of Seaham Harbour provides the stage for this novel. It was here that the aristocrat, the Marquess of Londonderry invested £165,000, a small fortune in 1828, to develop a new harbour that was to be the stimulus for the building of a new town. His grandson the 6th Marquess portrayed the typical flamboyant and fantastically wealthy playboy born into a privileged world but what secrets were lurking behind this façade?

How could the covert activities of this party loving nobleman have any bearing on the security of the nation?

Richard Raine is the master of the collier brigantine "The William Thrift" that enters Seaham Harbour in April 1881. Thus begins a tale that takes the reader by the hand and walks them through a time and lifestyle long forgotten. Richard could never have imagined the adventure that was to unfold when he left harbour and the part he would play in maintaining the fragile peace that existed in Queen Victoria's world. More than a century after these events took place the real story, "A Master Mariners Tale" is finally told.

SHIPBUILDING AT SEAHAM HARBOUR:

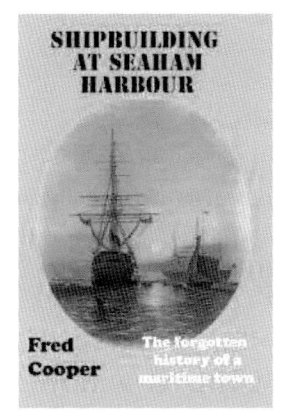

The town of Seaham stands proud on the Durham coast in the north-east of England. Seaham once built ships. Not large ships. Not iron ships - with one exception - but fine and sturdy wooden sailing ships. The shipyards are now gone; the patent slipways, pontoons, dry docks, workshops are all gone and no trace remains to remind Seaham of this once flourishing industry. The ships built at Seaham Harbour are here no longer. They either sank or were sent to the breakers yard many years ago. This book documents the fine history of shipbuilding at Seaham Harbour between 1832 and 1899.

HISTORY OF THE SEAHAM INFIRMARY

The 18th and 19th century was a torrid time for the sick, elderly and maimed in the United Kingdom. Care for the sick, injured, mentally ill and aged were provided from a variety of sources none of which was centrally planned or co-ordinated. The population of Seaham Harbour was rapidly expanding in the 1830's and 1840's with many new industrial and commercial ventures starting up and a rapidly expanding township. The time was right for some form of medical, surgical and nursing provision at Seaham Harbour. This is the definitive history of the Seaham Infirmary from 1844 to 1969.

A HISTORY OF THE LONDONDERRY LITERARY INSTITUTE, SEAHAM HARBOUR:

This book documents the definitive history of The Londonderry Institute at Seaham from it's opening in 1855. Built to serve the town as a Literary Institute it evolved into a focal point for public and social groups that bound the fabric of the community together. As it approached its centenary year it appeared to have reached the end of its useful life but the town would not let it be demolished.

SPORTING PASTIMES AT SEAHAM HARBOUR

The development of sport as a social pastime began to grow from the mid-19th century when working people began to enjoy the concept of the "weekend" when they did not work. This exploration of sporting pastimes at Seaham Harbour covers the period from 1835 through to the mid-20th century when sport and recreational activities were woven into the very social fabric of every-day life in the town.

A HISTORY OF THE CHURCHES AT SEAHAM

Churches are, for some, central to their spiritual existence and an important part of their day-to-day life. To some people visiting churches is a hobby. To others, churches are places they wander into whilst on holiday to look around, to sit and meditate in the quiet and to absorb the tranquillity of the occasion. Often the visitor's book is signed with comments such as beautiful, lovely or fascinating but do people really know anything about the building they have looked around? Do they know who built the church; why it was built; why it was erected in that location and at that time; what difficulties were encountered; where did the funding come from and what events or unique features set it apart from other churches? This book provides the answers to all of these questions about the twenty-four churches, past and present that were built in Seaham.

THE 2nd DURHAM (SEAHAM) ARTILLERY VOLUNTEERS

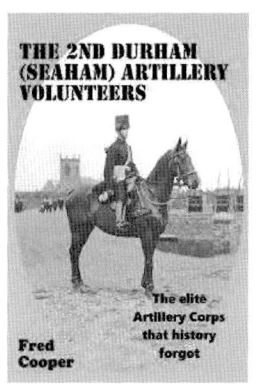

The 2nd Durham (Seaham) Artillery Volunteers were raised in 1860 at Seaham Harbour. More than 7,000 men from the colliery districts of Seaham, Silksworth, Rainton and Durham enrolled as members of the Corps. For three generations these men had a distinguished record amongst the Volunteer Artillery Brigades of Britain. This book is a record of the history and past service of the Old 2nd Durham's. Their descendants have a right to be proud of their record and achievement.

THE BOER WAR VOLUNTEERS FROM SEAHAM

This book provides the reader with the basic facts and details of events to explain the progress of the war from 1899 until its conclusion in 1902. However, the main purpose of the book is to record the contributions made by, and the real-life experiences, of men from Seaham during the conflict. The people of Seaham need to remember the courage, bravery and exploits of the Volunteers from their town who fought and died when their country called them to arms.

If you enjoy these books, please take a moment to leave a review for the book with your favourite eBook retailer. Why not visit my website at https://seahampast.co.uk and read more remarkable stories about Seaham.

Thank you,

Fred Cooper

NOTES